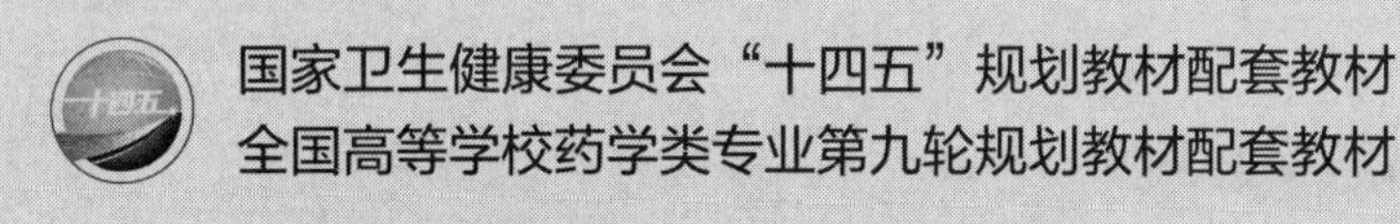

国家卫生健康委员会“十四五”规划教材配套教材
全国高等学校药学类专业第九轮规划教材配套教材

供药学类专业用

物理化学
实验指导（双语）

第4版

主　编　崔黎丽

副主编　袁　悦　王凯平　吴文娟

编　者　(以姓氏笔画为序)

王凯平(华中科技大学同济医学院)
刘　艳(北京大学医学部)
李　森(哈尔滨医科大学)
杨　峰(中国人民解放军海军军医大学)
吴文娟(广东药科大学)
陈　刚(复旦大学药学院)
林玉龙(河北医科大学)
袁　悦(沈阳药科大学)
栾玉霞(山东大学药学院)
崔黎丽(中国人民解放军海军军医大学)

人民卫生出版社

·北　京·

图书在版编目（CIP）数据

物理化学实验指导：英汉对照 / 崔黎丽主编 . —4 版 . —北京：人民卫生出版社，2023.9

ISBN 978-7-117-35200-0

Ⅰ. ①物… Ⅱ. ①崔… Ⅲ. ①物理化学 - 化学实验 - 高等学校 - 教学参考资料 - 汉、英 Ⅳ. ①O64-33

中国国家版本馆 CIP 数据核字（2023）第 162983 号

物理化学实验指导（双语）

Wuli Huaxue Shiyan Zhidao (Shuangyu)

第 4 版

主　　编：崔黎丽
出版发行：人民卫生出版社（中继线 010-59780011）
地　　址：北京市朝阳区潘家园南里 19 号
邮　　编：100021
E - mail：pmph @ pmph.com
购书热线：010-59787592　010 59787584　010-65264830
印　　刷：河北宝昌佳彩印刷有限公司
经　　销：新华书店
开　　本：787 × 1092　1/16　　印张：13
字　　数：324 千字
版　　次：2007 年 7 月第 1 版　　2023 年 9 月第 4 版
印　　次：2023 年 10 月第 1 次印刷
标准书号：ISBN 978-7-117-35200-0
定　　价：56.00 元
打击盗版举报电话：010-59787491　E-mail：WQ @ pmph.com
质量问题联系电话：010-59787234　E-mail：zhiliang @ pmph.com
数字融合服务电话：4001118166　E-mail：zengzhi @ pmph.com

前言

本书是全国高等学校药学类专业《物理化学》(第9版)的配套教材,采用双语体系编写,不仅可作为药学类院校学生的物理化学实验教材,还可供其他从事物理化学实验工作的有关人员参考,而且对于提高学生的科技英语水平有着积极的作用。

本书分为四个部分:第一部分为绪论。主要介绍物理化学实验的目的和要求、实验报告书写规则、实验室安全知识、误差理论和有效数字、实验数据的表示与处理,以及物理化学实验的设计思想,是成功完成实验的重要基础。第二部分为实验部分。在汲取参编院校长期实验教学经验的基础上,选编了热力学、相平衡、电化学、动力学、表面和胶体化学等物理化学分支中有代表性且较成熟的基本实验。这些实验既与课程内容紧密结合,又充分体现物理化学实验特点,突出基础性和实用性,同时注意融入绿色化学概念。此外,还编写了部分综合、设计性实验,在强化基础知识、基本技能训练的基础上,培养学生动手能力、创新思维能力和科学素养,培养学生分析问题和解决问题的能力。第三部分为实验测试题部分,为本版次的新增内容。选编了与基本实验相关的测试题,主要形式为选择题和简答题,希望在培养学生观察能力、实验技能和科学探究能力方面起到一定的强化作用。第四部分为附录部分。收录一些常用数据表以及本书编写过程中的参考文献,以便查阅。

双语编写体系是本书的一个重要特点,而突出药学特色是本书的又一重要特点。每个基础实验中都设有相关知识或实验技术的药学应用介绍,在上一版的基础上,本版次对这部分内容进行了充实和完善。设计性和综合设计性实验,内容涵盖药物多晶型、纳米材料的合成与表征、药物稳定性、药用表面活性剂和乳状液的性质,以及药物常规理化性质的测定,充分体现药学专业特色,提高了学生的学习兴趣和综合分析问题、解决问题的能力。

由于作者水平有限,书中难免存在问题和错误,恳请广大读者批评指正。

编者

2023年5月

Preface

This textbook is attached to the textbook of *Physical Chemistry*, edition ninth, and it is compiled in a bilingual system. It is suitable for readers majoring in medicine, pharmacy or other disciplines. Besides, it may help the students to learn and practice scientific English.

The textbook is divided into four parts. The first part contains introductions of purpose and requirement of experimental physical chemistry, report writing, laboratory safety, error theory, data processing, rules for experiment design, which are fundamental for students to finish the experiment successfully. The second part consists of a selection of some typical experiments in the areas of thermodynamics, phase equilibrium, electrochemistry, kinetics, surface and colloid chemistry. They are not only closely related to the knowledge that the students have learned in class, but also fully reflect the characteristics of physical chemistry experiment in fundamental, practicability, and concepts of green chemistry. Besides, some designing and comprehensive designing experiments are also developed to encourage ability in creative consciousness cultivation, problem solving and analyzing. The third part is experimental test questions, the newly added content in this edition, which consists of multiple choice and short questions related to twenty basic experiments to reinforce the ability in observation, experiment skills and scientific exploration. The fourth part is a list of reference data and books of experimental physical chemistry.

Bilingual system is an important feature of this book. Besides, to highlight pharmaceutical characteristics is another important feature of this book. The pharmaceutical applications of related principle and experimental technologies are included in each basic experiment, and are enriched and improved in this edition on the basis of previous edition. The designing experiments and comprehensive designing experiments in areas of polymorphism, nanomaterials, drug stability, properties of pharmaceutical surfactant and emulsion, as well as measurement of general physiochemical properties of drugs are compiled to reflect the pharmaceutical features, as well as to develop student's comprehensive and innovation skills. Therefore, this book is much more suitable for students majoring in pharmacy.

Due to limited abilities and inexperience, errors and negligence are inevitable in the textbook, and criticisms and advice are welcomed from our readers.

Editor

May 2023

目　录　（Contents）

第一部分　绪论 ……………………………………………… 1

一、物理化学实验的目的和要求 ……………………………… 1
二、物理化学实验的安全知识 ………………………………… 2
三、物理化学实验的设计思想和方法 ………………………… 4
四、误差理论和有效数字 ……………………………………… 4
五、实验数据的表示法 ………………………………………… 8

Part One　Introduction ……………………………………… 11

1. Objective and requirement …………………………………… 11
2. Physical chemistry laboratory safety ………………………… 12
3. Rules and methods for designing of the physical chemistry experiment …… 14
4. Error theory and significant figures ………………………… 15
5. Data reporting ……………………………………………… 19

第二部分　实验 ……………………………………………… 22
Part Two　Experiments ……………………………………… 22

实验一　燃烧热的测定 ………………………………………… 22
Experiment 1　Determination of Heats of Combustion ……… 26
实验二　溶解热曲线的测定 …………………………………… 29
Experiment 2　Determination of Heats of Solution Curve …… 34
实验三　凝固点降低法测定摩尔质量 ………………………… 39
Experiment 3　Determination of Molar Mass by Freezing
Point Depression …………………………………… 42
实验四　凝固点降低法测定氯化钠注射液的渗透压 ………… 45
Experiment 4　Determination of Osmotic Pressure of Sodium Chloride
Injection by Freezing Point Depression …………… 47
实验五　静态法测定液体饱和蒸气压 ………………………… 50
Experiment 5　Saturated Vapor Pressure of Pure Liquids by Static
Method ……………………………………………… 53
实验六　反应平衡常数及分配系数的测定 …………………… 56
Experiment 6　Determination of Equilibrium Constant of Reaction and

Partition Coefficient …… 58
实验七 二组分部分互溶双液系统相图的绘制 …… 60
Experiment 7 Drawing of Phase Diagram for a Partially Miscible Binary Liquid System …… 62
实验八 完全互溶双液系统平衡相图的绘制 …… 64
Experiment 8 Drawing of Phase Diagram for a Miscible Binary Liquid System …… 69
实验九 二组分简单低共熔系统相图的绘制 …… 74
Experiment 9 Drawing of Phase Diagram for a Binary Eutectic System …… 76
实验十 三组分液-液系统相图的绘制 …… 78
Experiment 10 Drawing of Phase Diagram for a Ternary Liquid System …… 81
实验十一 电解质水溶液电导的测定及应用 …… 83
Experiment 11 Measurement and Application of Electrolyte Solution Conductance …… 86
实验十二 电动势法测溶液 pH 和反应热力学函数 …… 89
Experiment 12 Determination of pH of Solutions and Thermodynamic Functions by Electromotive Force Measurements …… 93
实验十三 旋光法测定蔗糖转化反应的速率常数 …… 97
Experiment 13 Determination of Rate Constant for the Conversion of Sucrose by Polarimetric Method …… 103
实验十四 乙酸乙酯皂化反应速率常数及活化能的测定 …… 110
Experiment 14 Determination of Rate Constant and Activation Energy for the Saponification of Ethyl Acetate by Conductometric Method …… 113
实验十五 碘化钾与过氧化氢反应的速率常数及活化能的测定 …… 116
Experiment 15 Determination of Rate Constant and Activation Energy for Reaction between Potassium Iodide and Hydrogen Peroxide …… 119
实验十六 丙酮溴化反应速率常数的测定 …… 122
Experiment 16 Determination of Rate Constant for the Bromination of Acetone …… 124
实验十七 最大气泡压力法测定液体表面张力 …… 127
Experiment 17 Determination of the Solution Surface Tension by Bubble Pressure Method …… 131
实验十八 固体在溶液中的吸附 …… 134
Experiment 18 Adsorption of Solids in Solution …… 136
实验十九 溶胶的制备及性质 …… 139
Experiment 19 Preparation and Properties of Sols …… 142
实验二十 黏度法测定大分子的平均相对分子质量 …… 146
Experiment 20 Determination of Average Relative Molar Mass of the Macromolecule by Viscosity Method …… 150

实验二十一　药物多晶型的差热分析（设计性实验）…… 154
Experiment 21　Thermal Analysis of Polymorphism Pharmaceuticals（Designing Experiment）…… 155
实验二十二　药物稳定性及有效期测定（设计性实验）…… 156
Experiment 22　Determination of Drug Stability and Shelf Life（Designing experiment）…… 157
实验二十三　载药纳米粒子的制备及表征（综合设计性实验）…… 159
Experiment 23　Preparation and Characterization of Drug-loaded Nanoparticles（Comprehensive Designing Experiment）…… 160
实验二十四　固体药物常规理化常数的测定（综合设计性实验）…… 162
Experiment 24　Determination of Physical and Chemical Constants of Solid Drugs（Comprehensive Designing Experiment）…… 162
实验二十五　表面活性剂临界胶束浓度的测定（综合设计性试验）…… 163
Experiment 25　Determination of Critical Micelle Concentration for Surfactants（Comprehensive Designing Experiment）…… 165
实验二十六　乳状液的制备和性质（综合设计性实验）…… 166
Experiment 26　Preparation and Properties of Emulsion（Comprehensive Designing Experiment）…… 168

第三部分　实验测试题 …… 170

一、单选题 …… 170
二、简答题 …… 175
三、答案 …… 176

Part Three　Experiment Test Questions …… 179

Section One　Multiple Choice …… 179
Section Two　Short Questions …… 187
Section Three　Answers …… 189

第四部分　附录 …… 193
Part Four　Appendices …… 193

一、常用物理化学实验数据表（Common Data of Experimental Physical Chemistry）…… 193
二、参考文献（References）…… 199

第一部分　绪　论

物理化学实验是建立在化学各分支学科实验基础上的一门基础实验,综合了化学实验所需的基本研究工具和方法,在培养学生的基本实验技能、分析解决问题的能力和提高科研素质等方面占有特别重要的地位。

一、物理化学实验的目的和要求

物理化学实验的主要目的是使学生在巩固化学基本实验技能的基础上,初步了解物理化学的研究方法,掌握物理化学的基本实验技术和技能;巩固和加深对物理化学基本理论和基本概念的理解和掌握;能根据所学原理设计实验、选择和使用仪器;锻炼学生观察现象、获取实验数据、正确处理和分析实验结果的能力;培养学生严肃认真的科学态度和创新思维能力。

为了达到上述目标,必须做到以下基本要求。

1. 认真预习

(1) 实验前,应认真阅读实验教材及相关的参考书目和文献资料,明确实验目的和要求,掌握实验原理和方法。

(2) 了解仪器的结构和操作规程,明确实验内容和操作步骤。

(3) 根据对实验的理解,用简明扼要的方式写出预习报告,重点表述对实验原理和实验方法的理解,特别是实验操作步骤及操作过程中要注意的问题,并设计好记录原始数据的图表。

(4) 实验前,教师要检查每个学生的预习报告,必要时进行提问,并解答疑难问题。对未预习和未达到预习要求的学生,不得进行实验。

2. 严格、规范操作

(1) 进入实验室后,首先检查仪器和试剂是否符合要求,并做好实验前的各项准备工作。

(2) 在不了解仪器使用方法之前,不得擅自使用和拆卸仪器。仪器和线路安装或连接好后,须经教师检查无误后方能接通电源开始实验。

(3) 在教师指导下,严格按操作规程进行操作,不得随意更改。

(4) 仔细观察实验现象,如实、详细、准确地记录实验数据。要善于发现和解决实验中出现的问题。

(5) 实验结束后,应将实验数据交予指导教师审阅通过后,方能拆除实验装置。若不合格,则需重做或补做。

(6) 严格遵守实验室各项规则,保持实验室安静和整洁,尊重教师的指导。

3. 独立完成实验报告 实验后必须及时、认真地完成实验报告。实验报告必须独立完成,同一小组成员不得合写一份报告。实验报告要格式规范、内容完整、文字简练、表达清晰、结论明确,一般包括:①实验名称,实验日期,完成者姓名;②实验目的;③实验原理(简述);④实验内容,选用最简明扼要的方式表达每一项实验内容的操作步骤;⑤实验现象或实验数据;⑥实验结论、解释或实验数据处理、计算结果;⑦实验讨论,包括对实验中遇到的异常现象或问题的说明,实验结果的误差分析,实验的体会,或实验的改进意见等;⑧思考题。

实验报告不仅是概括实验过程和总结实验结果的重要的文献性资料,也是提高学生思维能力、专业能力和初步科研能力的重要的训练环节,必须高度重视。

二、物理化学实验的安全知识

物理化学实验中,潜藏着各种事故的危险。因此,每一个化学实验工作者必须具备一定的实验室安全防护知识。这里主要介绍安全用电、防火、使用化学药品和使用压缩空气的防护知识。

(一)安全用电防护

物理化学实验中大量使用电加热器、搅拌器、真空泵、各种电源及测量仪器等电器设备,如果不注意用电安全,将会导致触电和着火等事故,不仅危及实验者的生命,还将给国家财产造成巨大损失。因此,从安全防护出发,应做到以下几点:

(1) 不要用湿手或湿物接触通电设备。

(2) 使用前检查所有电器插头和电线绝缘情况,若有问题应及时更换。

(3) 电源的裸露部分应有绝缘装置(例如电线接头处应裹上绝缘胶布),所有电器的金属外壳应保护接地。

(4) 不要使超过规定负荷的电流流经电器,不要使电路过载。否则容易造成线路的过热,引起火灾和电击伤。

(5) 一般应以单手接触通电中的电器,将另一只手插入口袋或背在后面,以减小事故发生时电流流经胸腔的可能性,增加抢救的机会。

(6) 修理、安装电器,或实验前连接线路时,应先切断电源。实验结束后,先切断电源再拆线路。

(7) 不要在通电设备附近使用和放置易燃试剂,若有水或试剂洒落在电线或电器上,应拔去仪器插头或切断主电源。

(8) 若实验中仪器出现问题,不要自己修理,应及时报告老师,以免伤害自己或危及他人。

(9) 清楚了解实验室的安全设备和使用方法,特别要清楚紧急出口和灭火器的位置,以及实验室的电源总开关位置,一旦发生电线起火,便于及时拉开电闸,切断电源,再用一般方法灭火。若无法拉开电闸,可用沙土或 CO_2、CCl_4 灭火器灭火,禁止用水或泡沫灭火器等导电液体灭火。

(二)使用化学药品的安全防护

1. 一般安全防护知识 化学药品大多具有不同程度的毒性,毒物可以通过呼吸道、消化道和皮肤三种途径进入人体内。因此,为了尽量杜绝和减少毒物由上述途径进入体内,应做到以下几点:

(1) 实验前,应了解所用药品的毒性及防护措施。

（2）操作有毒气体或易挥发物质（如氰化物、高汞盐、有机溶剂等）应在通风橱中进行。可溶性钡盐、重金属盐（如镉、铅盐）、三氧化二砷等剧毒药品，应妥善保管，使用时要特别小心。

（3）使用可燃性气体时，要防止气体逸出，保持室内通风良好。同时严禁使用明火，还要防止产生电火花及其他撞击火花。

（4）使用有毒药品或可燃性、易挥发气体，注意自我防护，穿、戴相应的防护器具（如眼镜、手套、面罩等）。

（5）用移液管移取有毒、有腐蚀性的液体时，严禁用嘴吸。

（6）实验过程中，若有药品洒落或溅出，应戴好防透手套和护目镜立即清除。

（7）任何药品或试剂只能通过仔细阅读容器上的标签加以辨识，严禁舌尝或直接用鼻子闻。

（8）严禁将强酸和强碱或强氧化剂和强还原剂放在一起。

（9）化学药品用完后应倒入回收瓶（桶）中回收，不准倒入水槽中，以免造成污染。

（10）禁止在实验室内抽烟、喝水、吃东西。食品、饮料、香烟及化妆用品不要带进实验室，以防毒物污染，离开实验室及饭前要洗净双手。

2. 使用汞的安全防护　物理化学实验中接触汞的机会比较多，常温下汞蒸气容易溢出，吸入人体后将引起慢性中毒。汞蒸气的最大安全浓度为 $0.1mg/m^3$，而 20℃时，汞的饱和蒸气压为 0.16Pa，空气中的饱和浓度为 $15mg/m^3$，远远超过安全浓度。所以必须严格按照以下规定安全操作汞：

（1）汞不能直接露于空气中，在装有汞的容器中，汞面上应加水或其他液体覆盖。

（2）装汞的仪器下面一律放置塑料、瓷或不锈钢浅盘，一切转移汞的操作也应在装有水的浅盘中进行，防止操作过程中汞滴散落在桌上或地面上。

（3）万一有汞洒落在地上、桌上等地方，应首先打开窗户，并尽可能地用吸管将汞收集起来，再用能成汞齐的金属片（如 Zn、Cu）在汞溅落的地方多次扫过，最后用硫磺粉覆盖在有汞溅落的地方，使汞变成 HgS。不要用家用吸尘器吸取汞，也不要用笤帚扫。

（4）擦过汞或汞齐的滤纸或布必须放在有水的瓷缸内。

（5）盛汞器皿和有汞的仪器应远离热源，严禁把有汞仪器放进烘箱。

（6）切忌用有伤口的手接触汞。

（三）使用高压气体钢瓶的安全防护

高压气体钢瓶是物理化学实验中常用的仪器，如燃烧热测定等。使用高压气体钢瓶的主要危险是爆炸和气体泄漏（若是有毒气体或可燃性气体则更危险），因此，使用时应注意以下几点：

（1）使用前详细了解使用气体的性质、用途、安全防护方法。根据钢瓶外部标志和标签正确识别气体种类，不要把钢瓶颜色作为鉴定钢瓶内容物的主要手段，以免误用钢瓶。

（2）搬运及存放压缩气体钢瓶时，一定要将瓶上的安全帽旋紧。搬运充装有气体的钢瓶时，最好用特制的担架或小推车，也可以用手平抬或垂直转动。绝不允许用拉拽或滑动的方式或用手执着开关阀移动装有气体的钢瓶。

（3）高压气体钢瓶应贮存和使用于通风阴凉处，附近不得有还原性物质、热源、火种、电子线路。

（4）开启钢瓶的气门开关及减压阀时，应站在气阀接管的侧面，旋开速度不能太快，应逐渐打开，以免气体过急流出，发生危险。使用时先旋动开关阀，后开减压阀。用完，先关闭开关阀，放尽余气后，再关减压阀。不得只关减压阀，不关开关阀。

（5）钢瓶内气体不得全部用完，一定要保留 0.05MPa 以上的残留压力。

（6）绝不可使油或其他易燃性有机物沾在气瓶上（特别是气门嘴和减压阀）。也不得用棉、麻等物堵住，以防燃烧引起事故。

三、物理化学实验的设计思想和方法

物理化学实验在设计思想和方法上与一般的科学研究没有本质的区别。因此，学习和掌握物理化学实验的设计思路和方法可以使学生了解科学研究的一般过程，提升他们的批判性思维的技能，对创新思维和科研能力的培养十分有益。为此，本书安排了一些综合设计型实验。这里就实验的设计思想和一般步骤作一简单介绍。

实验方案设计时需注意科学性、安全性、可行性和简约性。实验方案的科学性是实验设计的首要原则，它是指实验原理、实验步骤和方法的正确性。安全性则要求实验设计时尽量避免使用有毒药品和有危险性的操作，以防造成环境污染和人身伤害。实验设计的可行性则是指实验设计切实可行，所选用的试剂、仪器设备在现有实验室条件下能够得到满足。而简约性是指实验简单易行，仪器简单易得，实验过程快速，实验现象明显。

设计性实验没有确定的研究体系，一般只给出一个方向，学生选定实验题目后，要根据题目中给定的信息，确定实验是对哪方面进行研究，利用课余时间查阅相关资料，在保证科学性的基础上，根据实验室现有的实验条件，运用所学知识，设计切实可行的实验方案及步骤，并写出实验设计报告。实验设计报告包括实验目的、原理、所需仪器、试剂、具体操作步骤、需记录的数据表格、数据处理方法等内容。实验设计报告应在实验前一周交给指导教师，经检查通过后方可按计划实施。实验中所需的试剂需自行配置，并根据仪器使用说明调试仪器，配置搭建实验装置，并独立完成整个实验操作，最后以论文形式给出实验报告。

设计实验开始时，往往是不完善的，可能会遇到很多困难。只有在实践中不断总结经验，不断修改完善，最后才能获得实验的成功。

四、误差理论和有效数字

在测量实验中，测量值和真实值不可能完全一致，其差值称为误差。分析测量结果的准确性和产生误差的主要原因，寻找减少误差的有效措施，可以提高测量结果的准确性。

（一）误差的分类

1. 系统误差　在同一条件下对同一量进行多次测量时，误差的符号保持恒定（即多次测量中均出现正误差或负误差，具有单一方向性），其数值按某一确定的规律变化，这种误差称为系统误差。

系统误差产生的原因包括仪器本身构造不完善而引起的误差、测量方法引起的误差、个人习惯误差和试剂误差等。

系统误差不能通过增加实验的次数使之消除，但通过改进实验方法、校正仪器、提高试剂纯度等，可以有针对性地使之减少到最小程度。

2. 偶然误差　偶然误差通常由一些不确定的因素所引起。从单次测量值看，误差的绝对值和符号的变化时大时小，时正时负，呈现随机性。但是其多次测量的结果服从概率统计规律。因此，可采用多次测量取算术平均值的方法来减小偶然误差对测量结果的影响，使测得结果接近真实值。

3. 过失误差 过失误差是不应出现的一些失误,如数据读错,试剂加错,记录、计算出错等。严格来说过失误差应是一种事故,不属于误差的范畴,也无规律可循。但是,只要实验者加强责任心,以严谨的科学态度进行操作,过失误差是可以避免的。

(二)准确度与精密度

准确度是指测量值与真实值符合的程度。若实验的准确度高,说明测量值与真实值越接近。精密度是指各测量值相互接近的程度。若所测数据相互十分接近,偏差很小,说明此实验结果的精密度高。

在分析测定过程中,由于存在误差且误差会传递,因而直接影响分析结果的精密度和准确度。系统误差仅影响分析结果的准确度,而偶然误差既影响精密度,也影响准确度。

评价分析结果应先看精密度再看准确度。但是,高精密度不能保证高准确度;而高准确度的数据却要足够的精密度来保证。因此,只有精密度和准确度都高的数值,才是可取的。

(三)实验误差的表示方法

误差的表示方法有多种,下面介绍常用的几种。

1. 绝对误差和相对误差 绝对误差是指测量值 x 与真实值 x_0 之差,相对误差是指误差在真实值中所占的百分数,分别表示为

$$\text{绝对误差}=x-x_0 \quad \text{式(1-1)}$$

$$\text{相对误差}=\frac{\text{绝对误差}}{\text{真实值}}\times 100\% \quad \text{式(1-2)}$$

用相对误差可以比较不同物理量的测量准确度。

2. 算术平均值与平均误差 在任何测量中,偶然误差总是存在。所以,我们不能以任何一次的观测值作为测量结果。为了使测量结果有较大的可靠性,常取多次测量的算术平均值。设对物理量 A 进行了 n 次测量,各次的测量值为 x_1、x_2、$x_3\cdots x_n$,则其算术平均值 $\bar{x}$ 为

$$\bar{x}=\frac{x_1+x_2+x_3+\cdots+x_n}{n} \quad \text{式(1-3)}$$

测定值与平均值之差称为测量的误差,用以衡量精密度的高低。其定义式为

$$\Delta x_i=x_i-\bar{x} \quad \text{式(1-4)}$$

Δx_i 值越小,测量的精度越高。又因为各次测量误差的数值可正可负,故需引入平均误差的概念,即

$$\overline{\Delta x}=\frac{|\Delta x_1|+|\Delta x_2|+|\Delta x_3|+\cdots+|\Delta x_n|}{n}=\frac{\sum_{i=1}^{n}|x_i-\bar{x}|}{n} \quad \text{式(1-5)}$$

而平均相对误差则为

$$\frac{\overline{\Delta x}}{\bar{x}}=\frac{|\Delta x_1|+|\Delta x_2|+|\Delta x_3|+\cdots+|\Delta x_n|}{n\bar{x}}\times 100\% \quad \text{式(1-6)}$$

3. 标准误差 若物理量 A 的个别测量值为 x_i,n 次测量的算术平均值为 $\bar{x}$,则标准误差 S 为

$$S=\sqrt{\frac{\sum_{i=1}^{n}(x_i-\bar{x})^2}{n-1}} \quad \text{式(1-7)}$$

用标准误差表示精密度要比用平均误差好，因为单次测量的误差平方之后，较大误差被更显著地反映出来，故更能说明数据的分散程度。

4. 间接测量的误差传递　物理化学实验中，很多实验结果不能直接测量（如物质的摩尔质量等），需通过其他可以直接测量的数据，经过数学运算间接得到，这种情况称为间接测量。这样，直接测量的误差将通过一定的规律传递到间接测量结果中，这就是间接测量的误差传递。研究误差传递的规律，可以帮助我们找到影响实验结果准确性的主要因素，从而有针对性地改进实验，确定实验条件和方法。

设物理量 U 是由直接测量值 x、y 经函数关系计算而得，即

$$U=f(x,y) \qquad 式(1\text{-}8)$$

若 Δx 和 Δy 分别为 x 和 y 的测量误差，且均比较小，则可以把它们看作微分 $\mathrm{d}x$ 和 $\mathrm{d}y$。因此，U 的测量误差 $\mathrm{d}U$ 为

$$\mathrm{d}U=\left(\frac{\partial U}{\partial x}\right)_y \mathrm{d}x+\left(\frac{\partial U}{\partial y}\right)_x \mathrm{d}y \qquad 式(1\text{-}9)$$

上式表明，各直接测量误差 $\mathrm{d}x$ 和 $\mathrm{d}y$ 都将影响最后结果 U，使其产生 $\mathrm{d}U$ 的误差。表 1-1 为误差在不同运算过程中的传递规律。

表 1-1　各种运算过程中的误差传递规律

Table 1-1　some examples for calculation of dU

函数式 function	绝对误差 absolute error	相对误差 relative error
$U=x+y$	$\pm(\lvert\mathrm{d}x\rvert+\lvert\mathrm{d}y\rvert)$	$\pm(\lvert\mathrm{d}x\rvert+\lvert\mathrm{d}y\rvert)/(x+y)$
$U=x-y$	$\pm(\lvert\mathrm{d}x\rvert+\lvert\mathrm{d}y\rvert)$	$\pm(\lvert\mathrm{d}x\rvert+\lvert\mathrm{d}y\rvert)/(x-y)$
$U=xy$	$\pm(x\lvert\mathrm{d}y\rvert+y\lvert\mathrm{d}x\rvert)$	$\pm(\lvert\mathrm{d}x\rvert/x+\lvert\mathrm{d}y\rvert/y)$
$U=\frac{x}{y}$	$\pm(x\lvert\mathrm{d}y\rvert+y\lvert\mathrm{d}x\rvert)/y^2$	$\pm(\lvert\mathrm{d}x\rvert/x+\lvert\mathrm{d}y\rvert/y)$
$U=x^n$	$\pm(nx^{n-1}\mathrm{d}x)$	$\pm(n\mathrm{d}x/x)$
$U=\ln x$	$\pm(\mathrm{d}x/x)$	$\pm(\mathrm{d}x/x\ln x)$

例　用冰点降低法测量溶质的摩尔质量时，所用计算公式为

$$M=\frac{1\ 000k_{\mathrm{f}}m}{m_0(t_0-t)}=\frac{1\ 000k_{\mathrm{f}}m}{m_0\theta}$$

式中，k_{f} 为冰点下降常数；m 为溶质质量；m_0 为溶剂质量；θ 为冰点下降度数；t_0 为溶剂冰点；t 为溶液冰点。由实验测出，$m=(0.300\ 0\pm0.000\ 2)$ g，$m_0=(20.00\pm0.02)$ g，$\theta=(0.300\pm0.008)$ ℃。求溶质摩尔质量的测量相对误差。

解：根据表 1-1 中误差传递的计算方法，摩尔质量的相对误差为

$$\begin{aligned}\frac{\Delta M}{M}&=\frac{\Delta m}{m}+\frac{\Delta m_0}{m_0}+\frac{\Delta\theta}{\theta}=\frac{0.000\ 2}{0.300\ 0}+\frac{0.02}{20.00}+\frac{0.008}{0.300}\\&=7\times10^{-4}+1\times10^{-3}+2.6\times10^{-2}\\&=0.028=2.8\%\end{aligned}$$

从以上计算结果可知，摩尔质量测量的精度完全取决于温度测量的精度，而称量的精度对结果的影响很微小。因此，要想减小测量误差，提高测量的准确度，应寻找更精密的测温

仪器或选用其他更好的实验方法，而不是采取过分准确的称量。

（四）有效数字

实验所获得的数值，不仅表示某个量的大小，还应反映测量这个量的准确程度。记录和计算测量结果都应与测量的误差相适应，不应超过测量的精确程度，即测量和计算所表示的数字位数，除末位数字为可疑者外，其余各位数从仪器上可直接测得。通常将所有确定的数字和末位不确定的数字一起称为有效数字。常用仪器的精度，如表 1-2 所示。

表 1-2 常用仪器的精度

Table 1-2 precision of some commonly used instruments

仪器名称（instrument）	仪器的精度（precision）	举例（example）	有效数字位数（number of significant figures）
托盘天平（scale）	0.1g	15.6g	3 位
1/100 天平（balance）	0.01g	15.61g	4 位
电光天平（photoelectric balance）	0.000 1g	15.606 8g	6 位
10ml 量筒（measuring cylinder）	0.1ml	8.5ml	2 位
100ml 量筒（measuring cylinder）	1ml	96ml	2 位
移液管（pipette）	0.01ml	25.00ml	4 位
滴定管（burette）	0.01ml	19.00ml	4 位
容量瓶（volumetric flask）	0.01ml	100.00ml	5 位

任何超出或低于仪器精度的数字都是不恰当的。例如，滴定管的最小刻度值为 0.1ml，读数记为 19.00ml。若将读数记作 19ml 或 19.000ml，前者降低了实验的精确度，后者则夸大了实验的精确度。

现列出有效数字的一些规则和概念：

（1）根据 0 在数字中的位置，确定其是否包括在有效数字的位数中。若 0 在数字的前面，只表示小数点的位置（起定位作用），不包括在有效数字中；若 0 在数字的中间或在小数点的末端，则表示一定的数值，应包括在有效数字的位数中。表 1-3 为一些含有 0 的有效数字的例子。

表 1-3 与“0”有关的有效数字位数确定

Table 1-3 determination of number of significant figures concerning zeroes

数值（data）	0.68	6.80×10^{-3}	0.023 50	6.08
有效数字位数（number of significant figures）	2	3	4	3

但是，整数 1 480 中的 0 无法明确是不是有效数字。为了避免这种困惑，常采用指数表示法。例如，137 000 表示三位有效数字，则可写成 1.37×10^5；若表示四位有效数字，则写成 1.370×10^5。

（2）对数值有效数字位数，仅由小数部分的位数决定，首数（整数部分）只起定位作用，不是有效数字。对数运算时，对数小数部分的有效数字位数应与相应的真数的有效数字位数相同。例如，pH = 7.68，其相应的真数为 $c(H^+) = 2.1\times10^{-8}$mol/L，即有效数字为两位，而不是三位。

（3）记录和计算结果所得的数值，均只能保留一位可疑数字。当有效数字的位数确定后，其余的尾数应根据“四舍六入五留双”的方法取舍。即当尾数≤4 时舍去，尾数≥6 时进位，当尾数恰为 5 时，则看保留下来的末位数是奇数还是偶数，若是奇数就将 5 进位，若是偶数则将 5 舍弃。总之应保留“偶数”，这样可以避免舍入后数字取平均值时又出现 5 而造成系统误差。例如，将 2.123 4 和 2.123 6 变为四位数时应分别为 2.123 和 2.124，将 3.123 5 和 3.124 5 变为四位数时均为 3.124。

（4）若第一位的数值≥8，则有效数字的位数可以多算一位。例如 9.15，虽然实际上只有三位有效数字，但在运算时可以看作四位有效数字。

（5）加减法运算时，各数值小数点后面所取的位数与其中最少者相同。乘除法运算时，所得的积或商的有效数字，应与各值中有效数字位数最少者相同。

五、实验数据的表示法

物理化学实验数据的表达方式主要有列表法、作图法和方程式法。

（一）列表法

实验结束后，将测得的一系列数据按自变量和因变量的对应关系用表格列出，这种表达方式称为列表法。列表法简单易行，便于参考比较，实验的原始数据记录一般采用列表法。使用列表法时应注意以下几点。

（1）表格应有简明而又完整的名称。

（2）每行或每列的第一栏应标出变量的名称及单位。

（3）表中的数据应化为最简单的形式，公共的乘方因子应在第一栏的名称中注明。

（4）每行中数值的排列要整齐，位数和小数点要对齐。

（5）实验条件和环境条件应在表中或表外注明，如室温、大气压、测定日期和时间等。

（二）作图法

1. 作图法的应用　用几何图形来表示实验数据的方法称为作图法。作图法有很多优点，如能清晰显示数据的变化规律；能直观看出数据之间所显示的特点，如直线、曲线、极大、极小和转折点等；能利用直线求斜率，由曲线求切线；还能用内插、外推等方法对数据做进一步处理。

作图法在物理化学实验中主要有以下几方面的应用：

（1）求内插值：根据实验所得数据，作出函数间相互关系的曲线，然后找出与某函数相应的物理量的数值。例如，完全互溶双液系相图绘制实验中，首先作出环己烷-乙醇系列标准溶液的折光率-组成工作曲线，然后根据所测系统的折光率，由工作曲线求出所测系统的组成。

（2）求外推值：在某种情况下，测量数据间的线性关系可以外推到测量范围之外，求某一函数的极限值，这种方法称为外推法。很多情况下，外推法可以推广到无法用实验方法测量的范围中。例如，黏度法测定高聚物分子量时，需求得特性黏度。虽该值不能由实验直接测定，但可先测定不同浓度时的比浓黏度，再作图外推至浓度为 0 处，即得特性黏度。

（3）求任何一点函数的导数：过曲线上的已知点作切线，求出切线的斜率即为该点函数的导数，是物理化学实验数据处理中常用的方法。例如，以反应物浓度对时间作图，在不同时间下求曲线切线的斜率即为该时间的反应速率。

(4) 求经验方程式:若函数和自变量有线性关系:$y=mx+b$,则以相应的 x 和 y 的实验值作图,得到一条尽可能连接各实验点的直线,由直线的斜率和截距可求出方程式中 m 和 b 的数值,代入上述方程即得所求经验方程。例如,固体在溶液中的吸附实验中,吸附量和平衡浓度的关系为:

$$\frac{x}{m}=Kc^n$$

对上式两边取对数,转换为线性关系:

$$\lg\frac{x}{m}=\lg K+n\lg c$$

以 $\lg\frac{x}{m}$ 对 $\lg c$ 作图,式中的经验常数 K 和 n 可由直线的斜率和截距分别求得。

2. 作图法基本要点 掌握娴熟的作图技术对于获得优良实验结果至关重要,下面为作图法的基本要点。

(1) 坐标纸和比例尺的选择:直角坐标纸最常用,有时也用单对数坐标纸或双对数坐标纸,在表达三组分系统相图时,常用三角坐标纸。

在直角坐标图上,习惯用横坐标表示自变量,以纵坐标表示因变量,横、纵坐标读数不一定从零开始,但应充分合理地利用坐标纸的全部面积。

为了能从图上迅速读出任一点的坐标值,坐标分度宜选 1、2、5、10 的倍数。直角坐标的两个变量的全部变化范围在两个坐标轴上表示的长度要相近,否则图形会扁平或细长。若所作图形为直线,则两坐标轴标度的选择应使直线的斜率值在 1 附近。

(2) 实验点的描绘及连线:习惯上用 •、○、Δ 等符号表示实验点。在同一张图纸上,不同的物理量应选用不同的符号表示,以示区别,并在图中注明。

作曲线时,应根据所描的数据点,将曲线光滑、连续地描出。通常曲线并不能通过所有数据点,应使数据点平均地分布在曲线两旁,或使所有的实验点离开曲线距离的平方和最小,此即“最小二乘法原理”。

(3)图名与说明:每个图都应写上简明的图题、横纵坐标表示的物理量名称、标度和单位,以及主要的测定条件如温度、压力和测定日期等。横纵坐标的标注应是纯数,物理量和单位之间用斜线“/”隔开。

用计算机作图给数据处理带来了极大的方便,但是应用计算机作图时,也要遵循以上规则。

(三)方程式法

用数学方程式表示实验数据的方法称为方程式法。该法不但表达方式简单,记录方便,而且能在实验范围内计算与自变量相对应的函数值,并能对所得方程式进行微分、积分和内插求值。

通常情况下,两个变量间的关系是已知的。但是,当两个变量间存在的具体关系未知时,可以先作图,由图形的形状与已知方程式相对应的图形比较,判断曲线的类型。由于直线关系式最简单而又容易直接检验,因此,对所得的函数关系式要尽量通过函数变化将其直线化,用图解法求出该直线的斜率和截距,即直线方程式 $y=a+bx$ 中的 a 和 b 两常数。但是,在很多情况下,变量之间的关系为 $y=a+bx+cx^2+dx^3+\cdots$ 的多项式,此时可对数据进行数学拟合,求各常数项,而多项式项数的多少以结果满足实验误差要求为准。

直线方程在物理化学实验中十分重要，通常可用作图法、平均值法和最小二乘法确定方程中的常数 a 和 b。其中最小二乘法处理较复杂，但结果可靠。若根据最小二乘法原理，用计算机编制程序，不仅可以快速简捷、准确地给出直线方程中的常数，而且还能给出相关系数。

Part One Introduction

Experimental physical chemistry is of great importance in training the basic experimentation and encouraging ability in problem solving and research.

1. Objective and requirement

The purpose of the experimental physical chemistry is to reinforce the skills in fundamental experiment methodology in other chemistry laboratory courses, to train the student in the research methods and techniques of physical chemistry, to amplify and make more meaningful of the abstract concepts, principles and equations treated in the textbook, to train the student in the processes of obtaining, processing and interpreting experimental data, and to encourage ability in research.

To achieve above objectives, there are some remarks to students as follows.

1.1. Read the upcoming experiment carefully

(1) Read the experimental text book and references carefully and thoroughly before entering the laboratory. Familiarize yourself with the objective, requirement, principle and technique of the experiment.

(2) Familiarize yourself with the structure and operation method of the instrument. Make a good understanding of the experimental content and procedures.

(3) According to your advance reading, write down briefly the principle and procedures, especially questions concerning the experiment, draw in your laboratory notebook columns or tables that will allow you to record the data.

(4) To ensure you have read the upcoming experiment, the instructor will check your notebook and ask or answer the questions concerning the experiment. If you fail to complete the pre review, you will not be allowed to perform the experiment.

1.2. Do the experiment seriously

(1) Make sure that the instruments and chemicals are of experimental requirement after entering the laboratory, otherwise change them with your instructor. Get everything ready promptly before starting the experiment.

(2) Do not use the instruments or unload the apparatus before a well understanding of them. Do not start the experiment unless the circuit connection or apparatus setting is examined by your instructor.

(3) Do the experiment as instructed. Never change the procedure at will unless otherwise instructed to do so by the instructor.

(4) Always observe carefully during the experiment, and record all experimental data as they

are read. It is encouraged to find problems and solve them during the experiment.

(5) When all the necessary measurements have been completed, do not dismantle the equipment before handing in your notebook to have your instructor sign on the data sheet. If the measurement is not qualified, repeat the experiment.

(6) Obey the laboratory policy strictly, keep quiet in laboratory and have high regard to instructor's guidance.

1.3. Write the report independently

After finishing the experiment, make suitable data processing and write the experiment report independently. A shared report is not allowed among the group members. A good report should be clear, concise and well organized. Usually, the report for each experiment will consist of the following parts: ①title of the experiment, the date on which the experiment is done, and the name of the accomplisher; ②experiment objectives; ③brief introduction of principle; ④a description of experiment procedure; ⑤experiment phenomena or data; ⑥results, explanation and calculations; ⑦a discussion, especially the explanation of the abnormal phenomena observed during the experiment, cause of error analysis, and some suggestions for the experiment; ⑧key to the questions.

Experiment report is not only the summary of the experiment procedure and results, but also an important way to encourage the student's ability in thoughts, knowledge and research.

2. Physical chemistry laboratory safety

During the course of experimental physical chemistry, there is always the potential to expose variety of possible hazards. Therefore, it is essential for all persons to be informed and guided in safe practices that should help to avoid injury. In this part, some safety practices and guidance concerning electricity, fire, chemicals and compressed gases are introduced.

2.1. Electrical safety

Electrically powered equipments, such as heater, stirrer, vacuum pump and varies of power supplies and instruments etc., are essential elements of physical chemistry laboratory and pose a significant hazard to laboratory workers and properties, particularly when mishandled or not maintained. The major hazards associated with electricity are electrocution and fire. To protect from the hazards caused by electricity, one should follow some basic precautions.

(1) Avoid contacting equipment that is plugged in with wet hands or wet materials.

(2) Inspect all electrical equipment before use to ensure that cords and plugs are in good condition. Remove from service or get repaired immediately if damaged or defective cords or plugs are found out.

(3) All exposed electrical conductors must be behind shields. All electrical equipment must be in proper grounding.

(4) Do not overload circuits or wiring. Overloading can lead to overheated wires and arcing which can cause fires and electrical shock injures.

(5) Use one hand to contact the operating equipment, and keep the other hand behind your back or in your pocket to reduce the likelihood of accidents that result in current passing through the chest cavity.

(6) Do turn off the power source before repairing or assembling electrical equipment, or

connecting the experimental circuit. Do turn off the power before disconnecting circuit when the experiment is finished.

(7) Do not use or store highly flammable solvent near equipment that is plugged in. Unplug the equipment or shut off the main power if water or a chemical is spilled onto equipment.

(8) If a piece of equipment fails while being used, report it immediately to your instructor. Never try to fix the problem yourself because you could harm yourself and others.

(9) Know the location of the lab safety equipment and understand how to use it. In particular, know the location of the emergency exit, fire extinguisher. Know clearly the location of main electrical switch in your laboratory. Turn off the power immediately in the event of a fire or electrocution. Do not use water or foam on an electrical fire. The appropriated fire extinguisher is sand, or CO_2 and CCl_4.

2.2. Chemical safety

2.2.1. General rules of safety: Most of the chemicals are potentially hazardous. Exposure to hazardous chemicals may occur by inhalation, ingestion and skin adsorption. Therefore, it is important for the students to be educated in general chemical safety and to follow some precautions as below.

(1) Identify chemical as well as review and understand hazardous and protective procedure before use.

(2) Operations of hazards and noxious, volatile chemicals such as cyanide, mercury salts and organic solvent should be done only in a laboratory hood. Special care should be taken when storage and process of toxic chemicals such as soluble barium salt, heavy metal salts and arsenic trioxide are concerned.

(3) Maintain adequate ventilation when using flammable gases. Keep away from heat, spark, and sources of ignition.

(4) Wear adequate gloves, glasses or face mark when process toxic chemicals, or volatile, flammable gases.

(5) Use pipette to remove the toxic or corrosive liquid. Never pipette by mouth.

(6) Clean up spills immediately, wearing impervious gloves and safety glasses.

(7) Read the label on chemical bottle to identify its content. Do not taste, or smell chemicals.

(8) Do not store acids and bases together. Do not store strong reducers and oxidizers together.

(9) Dispose of waste chemicals in proper containers. Never pour them down the drainpipe.

(10) Never eat, drink, or smoke while working in the laboratory. No food, beverage, tobacco, or cosmetic products are allowed in the laboratory to avoid toxic contamination. Wash hands before leaving the laboratory and before eating.

2.2.2. Mercury safety: There is great potential for you to use mercury containing devices in physical chemistry laboratory. Mercury is an extremely volatile and toxic chemical. Inhalation of mercury vapor may result in chronic poisoning. The maximum safety concentration of mercury vapor is $0.1 mg/m^3$. However, the liquid mercury has a vapor pressure of 0.16Pa and a saturated concentration of $15 mg/m^3$ in the air at 20℃. Therefore, it is fairly important to follow the practices when using mercury.

(1) Do not leave open containers of mercury in the laboratory. Add a layer of water or other liquid over the mercury in the container to avoid direct exposure of mercury in the air.

(2) Equipments containing mercury or handling mercury should have steel, porcelain, or plastic tray around the container in case of mercury spilling.

(3) For mercury spills, immediately open windows. Use pipette, syringe to pick up the mercury, and then use mercury absorbing metal such as Zn or Cu to create an amalgam, finally apply powdered sulfur on an affected surface to bind with mercury to HgS. Never use a house vacuum cleaner or a broom to clean up mercury spills.

(4) After cleaning up the mercury, place the used filter paper or cloth in aporcelain container containing water.

(5) Keep the mercury containing instrument and vessel away from heater. Never put the mercury containing instrument into the oven.

(6) Never contact mercury with a cut or wound hand.

2.3. Compressed gas safety

Compressed gas cylinders are routinely used in physical chemistry laboratory. The potential hazards associated with compressed gases include explosion, gas leakage (much more dangerous for toxic and flammable gases). To prevent the hazards, there are several general procedures to follow for safe handling and use of a compressed gas cylinder.

(1) Know and understand the properties, uses and safety precautions of the gas before using the cylinder. The contents of any compressed gas cylinder should be identified clearly by labeling or marking on the gas cylinder. Color coding is not a reliable means of identification.

(2) Cylinders should be transported with safety cap being screwed and fitted securely. Always use a cylinder cart to move compressed gas cylinder. Do not move the cylinder by sliding, rolling, dragging, or by the valve.

(3) Cylinders should be stored and used in a well ventilated area away from reducers, flames, source of heat or ignition, electric circuits.

(4) Open cylinder valves and regulator slowly to avoid a sudden release of the gases that will lead to a dangerous situation. Stand to the side of the regulator when opening the cylinder valve. Open the cylinder valve before the pressure reduce regulator is opened when the system is in use. Close the cylinder valve and release all pressure from the regulator before a regulator is removed from a cylinder when a system is not in use.

(5) A cylinder should never be emptied to a pressure lower than 0.05Mpa.

(6) Do not permit oil or flammable organics to come in contact with cylinder or their valves. Never use cotton or linen for leakage protection on the regulator and cylinder valve.

3. Rules and methods for designing of the physical chemistry experiment

There is no basic difference between physical chemistry experiment and science research in design rules and methods. Thus, developing the ability to design an experiment and understanding some considerations in experiment designing is critical to comprehending the scientific process and in promoting critical thinking skills. This skill is necessary for training in abilities of innovation thought and scientific research. To meet the objectives, some designing experiments are arranged in this book.

Science, safety, feasibility and parsimony are important in experiment designing. Science refers to the correctness of experiment principle, procedures and methodologies. Safety refers to avoiding use of noxious substances and dangerous operation to prevent from surrounding contamination and personal injury. The feasibility of experiment design means that all the reagents and instrument you chosen should satisfy the practical constraints of the laboratory condition. And then, parsimony means the simplest instrument and experiment route, the quick measurement process and obvious experiment phenomena.

Generally, the designing experiments only give you a topic and have no specific research system. The students have to read appropriate journal or literature related to your experiment to construct the experiment according to the laboratory conditions and write an outline of experiment design for instructor's evaluation one week ahead of the experiment. The outline should include experiment objective, principle, instruments required, reagents, procedures, data recording table and data process method etc. Modify the experiment according to instructor's evaluates or suggestions. Once the experiment has been approved, prepare all the reagents, adjust the instruments, install apparatus, run the experiment and write a final report of experiment in form of thesis.

The initial experiment design is usually imperfect and you may encounter many difficulties. A good way to improve procedure is to conduct the experiment. For each experimental run, you may determine the optimum experimental condition and procedures. Then you can obtain a satisfying result.

4. Error theory and significant figures

We can never measure the true value of any quantity but only an approximation to it. The difference between the observed value and true value is called the error of a measurement. To analyze the precision of the measurement and the cause of the errors can help us to take the steps in error reducing, and to improve the measurement precision.

4.1. Types of errors

4.1.1. Systemic errors: Systemic errors refer to the errors which persist during a series of experiments of a given type, i.e., the value of the measurement will always be greater (or less) than the true value. Systemic errors may be caused by defects in the instrument, using improper experiment design, impurity of chemicals, or sometimes imperfect usage of the instrument by the experimenter. Systemic errors cannot, in general, be eliminated by any process of averaging, but can be avoided by methodological improvement, proper calibration and adjustment of the instruments, and upgrading the chemical purity and so on.

4.1.2. Random errors: Random errors usually arise from some indeterminate causes and manifest themselves in variations in successive readings of a magnitude by the same observer under the same set of conditions. However, they can be treated through statistical analysis and can be reduced by averaging over a large number of readings or observations.

4.1.3. Mistakes: Mistake is an error which can be traced to carelessness, inattention, and various negative attitudes. Mistakes can be reduced in most cases by developing the experimenter in a way that he or she learns to be more careful, attentive, conscientious and proud of the work being done.

4.2. Precision and accuracy

Accuracy is the degree of conformity with a true value. The higher the accuracy is, the less the difference between the measured value and true value will be. Precision is the degree of reproducibility of the measurement. If a little divergence among the readings is shown, the method of measurement is one of high precision.

The measurement of a physical quantity can never be made with perfect accuracy and precision, because of some errors or uncertainty present, and because of the propagation of errors. In general, systemic errors may only affect the accuracy of the measurement. On the other hand, random error may affect both the accuracy and precision.

A precise measurement is not necessarily accurate, though we should concern the precision first other than accuracy in judging the quality of the measurement. However, an accurate measurement is guaranteed by a high degree of precision. A result of both precision and accuracy is valuable.

4.3. Error analysis

4.3.1. Absolute error and relative error: Absolute error is the difference between the measured quantity x and the true value x_0, given by

$$\text{Absolute error} = x - x_0 \tag{1-1}$$

Relative error is a ratio of absolute error to actual value, given by

$$\text{Relative error} = \frac{x - x_0}{x_0} \times 100\% \tag{1-2}$$

Relative error helps us to compare how close the quantity is from the value considered to be true.

4.3.2. Arithmetic mean and average error: The existence of random errors makes the result of a single measurement be not reliable in principle. Only with frequent repetition of the measurement do we get the validity of a measurement result. If x_1、x_2、$x_3 \cdots x_n$ is a series of n measurements obtained independently under identical experimental condition, the arithmetic mean or average value $\bar{x}$ of the measurements is defined as

$$\bar{x} = \frac{x_1 + x_2 + x_3 + \cdots + x_n}{n} \tag{1-3}$$

The difference between measured value and average value is the deviation which represents the precision of the measurement and can be expressed as

$$\Delta x_i = x_i - \bar{x} \tag{1-4}$$

The smaller the deviation is, the more precision the measurement is. Since the magnitude of the random errors is different for every repetition of the experiment, average error is defined and can be calculated using the equation

$$\overline{\Delta x} = \frac{|\Delta x_1| + |\Delta x_2| + |\Delta x_3| + \cdots + |\Delta x_n|}{n} = \frac{\sum_{i=1}^{n} |x_i - \bar{x}|}{n} \tag{1-5}$$

Then the percent relative error is given by

$$\frac{\overline{\Delta x}}{\bar{x}} = \frac{|\Delta x_1| + |\Delta x_2| + |\Delta x_3| + \cdots + |\Delta x_n|}{n\bar{x}} \times 100\% \tag{1-6}$$

4.3.3. Standard deviation: The standard deviation is a measure of the dispersion of a set of data from its mean and is given by

$$S = \sqrt{\frac{\sum_{i=1}^{n} (x_i - \bar{x})^2}{n - 1}} \tag{1-7}$$

Standard deviation is also typically called the uncertainty in a measured result. It is more reasonable to compare the dispersion degree of the data using standard deviation than using mean error because squaring each difference makes the bigger differences stand out.

4.3.4. Propagation of errors: Some quantities, such as molar mass of substance, cannot be measured directly through experiment, but you can calculate it depending on some other quantities which you can measure. This is called indirect measurements. Then errors from direct measurements may have effects on result of the indirect measurement. This is defined as the propagation of errors. Understanding of the rule of error propagation can help us to find the main factors which affect the experiment result, to improve the experiment accordingly, and to create appropriate experiment condition and method.

Suppose the numerical property U is the function of x and y

$$U = f(x, y) \tag{1-8}$$

The measurement of quantities x and y yields uncertainties of Δx and Δy. If the errors are small, then Δx and Δy can be taken as dx and dy. The total differential of U is given by

$$\mathrm{d}U = \left(\frac{\partial U}{\partial x}\right)_y \mathrm{d}x + \left(\frac{\partial U}{\partial y}\right)_x \mathrm{d}y \tag{1-9}$$

Therefore, the errors for x and y will have an effect dU on the calculation quantity U. Several examples for calculation of dU are listed in Table 1-1.

Example. The molar mass of the solute can be measured by the depression of freezing point method and is calculated using the equation

$$M = \frac{1\,000 k_{\mathrm{f}} m}{m_0 (t_0 - t)} = \frac{1\,000 k_{\mathrm{f}} m}{m_0 \theta}$$

where k_{f} is the constant of depression of freezing point, m is the mass of solute, m_0 the mass of solvent, θ is the degree of freezing point depression, t_0 is the freezing point of solute, and t is the freezing point of solution. The measurements give that $m = (0.300\,0 \pm 0.000\,2)$ g, $m_0 = (20.00 \pm 0.02)$ g, $\theta = (0.300 \pm 0.008)$ ℃. Please calculate the relative errors for determination of molar mass of the solute.

Solution. According to table 1-1, we have

$$\frac{\Delta M}{M} = \frac{\Delta m}{m} + \frac{\Delta m_0}{m_0} + \frac{\Delta \theta}{\theta} = \frac{0.000\,2}{0.300\,0} + \frac{0.02}{20.00} + \frac{0.008}{0.300}$$

$$=7\times10^{-4}+1\times10^{-3}+2.6\times10^{-2}$$
$$=0.028=2.8\%$$

From the above calculation we know that the precision for molar mass determination mainly depends on the precision of temperature determination, and accurate mass measurement affects a little on the experiment result. Therefore, the best way for error reducing and precision improvement is using more precision apparatus for temperature measurement or using other better experiment method instead of paying too much attention to accurate mass measurement.

4.4. Significant figures

The data obtained in measurement reflect both the magnitude of the quantity and the degree of accuracy. It means that digits recorded or reported should be related only to the precision of the observation. The number of digits expressed in a measurement should be read directly from the instrument except for the end doubtful one. Significant figure often been defined as "those that are read directly, plus the first doubtful one". The precision of some commonly used instruments are listed in Table 1-2.

Any figure that under or beyond the precision of instrument is considered inappropriate. A recorded measurement viewed with burette as 19.00ml does not mean the same thing as 19ml or 19.000ml. The former lowers the precision of the experiment, while the latter enlarges the precision of the experiment.

There are some rules in dealing with the significant figures.

(1) Whether zeroes are included in number of significant figures or not depends on their positions in the figure. Zeroes at the beginning (either to left or right of the decimal point) are not significant. Such zeroes merely indicate the position of the decimal point. Any zeroes between two significant digits or a final zero in the decimal portion only are significant. Table 1-3 are examples of determination of number of significant figures with the zeroes.

However, when a figure ends in zeroes that are not to the right of a decimal point, the zeroes are not necessarily significant. For example, 1 480 may be 3 or 4 significant figures. The potential ambiguity in this situation can be avoided by the use of scientific notation. We can write 137 000 as 1.37×10^5 (3 significant figures), or 1.370×10^5 (4 significant figures).

(2) The number of significant figures of a logarithm figure is determined by the decimal portion only. The figure at the beginning (integer portion) serves only as space holder. When you do the calculation, the digit of significant figure in the decimal portion should equal to that of the exact number. For example, pH = 7.68, then the exact number is $c(H^+)=2.1\times10^{-8}$ mol/L. It has two significant figures, but not three.

(3) The last digit of reported or calculated data is retained by the first doubtful digit. Then four round six into five double. That is, when the mantissa $\leqslant 4$ is rounded off, and when the mantissa $\geqslant 6$ is carried. If the digit to be dropped is greater than 5, the last retained digit is increased by one. If the digit to be dropped is less than 5, the last retained digit is left as it is. If the digit to be dropped happens to be a 5, an arbitrary rule can be used. Increase by one, if the digit before the 5 is odd. However, let it be, if the digit before the 5 is even. For example, when round to 4 significant digits, 2.123 4 and 2.123 6 become 2.123 and 2.124 respectively, but 3.123 5

and 3.124 5 both become 3.124.

(4) If the leading digit is equal to or greater than 8, increase the number of significant figure by one. For example, we can take the number of 9.15 as 4 significant figures in calculation.

(5) In addition and subtraction, retain as many decimal figures as are given in the number having the fewest decimals. In multiplication and division, the result should be retained to have the same number of significant figures as in the components with the least number of significant figures.

5. Data reporting

Most commonly, the data obtained in physical chemistry are presented in the form of tables, graphs or equations.

5.1. Tables

After experiment, organize the data observed into a table according to the relationship between independent variables and dependent variables. It is easy to create a table and make comparison among a list of data. Theraw data are commonly presented in a table. There are some guidelines for creating a good table.

(1) The table should have a title that provides a short description of the table's purpose.

(2) The columns and rows should be labeled and the physical name and units must be indicated.

(3) The data reported in table should be in their simplest form. If a same power of ten is used, it should be indicated in the column heading.

(4) The decimal point of the numbers in a column should be aligned.

(5) In general, the condition of experiment or surroundings should be specified in table, such as room temperature, atmospheric pressure, date and time of the measurement.

5.2. Graphs

5.2.1. Use of graphs: The use of graphs in presenting experiment data is called graphic methods. Graphing techniques are very useful in experimental physical chemistry mainly to (a) give a pictorial presentation of the variance regularity of the data; (b) reveal maxima, minima, inflection points, or other significant features; (c) perform direct differentiation by drawing tangents to a curve, accomplish integration, and so on. Some examples of uses of graphs are as follows.

(1) Interpolation method

Use the raw data from the experiment to plot the graph. The curve shows the relationship between dependent variable and independent variable. An intermediate value of one variable (dependent or independent), can be obtained from the curve. For example, in drawing a phase diagram of a binary liquid system of cyclohexane and ethanol, the compositions of vapor and liquid at equilibrium are obtained by interpolation method.

(2) Extrapolation

Under certain circumstance, the data of interest is beyond the scope of measurement. In such case the line in the graph may be extrapolated to obtain the required value. For example, in the determination of the molar mass of a polymer by viscosity method, the intrinsic viscosity (viscosity of the solution at infinite dilution) necessary for the calculation of the molar mass is derived by extrapolation to zero concentration.

(3) Graphic differentiation

Select several data points on the line and draw tangents at the points. Calculate the slopes of the tangents, given the derivative. Graphic differentiation enables us to find the reaction rate by plot the reactant concentration against time. The slope of the tangents on the line at different time is the reaction rate.

(4) Acquiring empirical equation

If the experimental measurement of the quantities x and y follows the equation: $y = mx + b$, plot x versus y, a straight line can be obtained. Find out the slope m and the intercept b of the line. Substitute the value of m and b into the equation to obtain the analytical equation. For example, the relationship between adsorption amount and concentration at adsorption equilibrium follows

$$\frac{X}{m} = Kc^n$$

Taking logarithm of both sides of the above equation gives

$$\lg\frac{x}{m} = \lg K + n\lg c$$

Then a straight line can be obtained by plotting $\lg\frac{x}{m}$ versus $\lg c$. The constants K and n can be calculated from the slope and intercept.

5.2.2. Graphing guidelines: There are also some important guidelines for making a good graph.

(1) Selecting graph paper and scale

The most commonly used graph paper is ordinary rectangular coordinate paper. Semilogarithmic paper is convenient when one of the coordinate is to be the logarithm of an observed variable. If both coordinates are to be the logarithms of variables, log-log paper may be used. Another special purpose paper which has triangular coordinates is used to plot the ternary phase diagram.

The independent variables must be plotted along the abscissa and the dependent variables along the ordinate. The beginning scale of the coordinate is not necessarily numbered zero. The scales should be numbered so that the resultant curve is as extensive as the sheet permits.

The scale number should be multiples of 1, 2, 5 or 10 for easy reading. Scales should be chosen such that the curve will, to the extent possible, have a geometrical slope approximating unity.

(2) Plotting the data and fitting the curve to the plotted points

Each point should be plotted with a suitable symbol, such as a dot "•", circle "○", triangle "Δ" etc. If several curves are plotted on the same sheet of graph paper, a different type of symbols should be used for each set of data.

After putting down all the data points on the graph, a smooth curve is drawn through the points. And the curve should pass as close as reasonably possible to all the plotted points, or the sum of squares of the difference between experimental value and the value predicted by a liner equation is the least, that is the principle of least squares.

(3) Preparing a descriptive title

A graph should include a complete descriptive title of what the graph is intended to show, the name of the physical quantity, scale and unit. The number of the scale should be pure figure. A bias "/" is used to separate the physical property and its unit.

The above rules should also be followed while using computer to plot the graph, though it is much convenient to manipulate the data.

5.3. Equation

In order to obtain the maximum usefulness from a set of experimental data, it is frequently desirable to express the data by a mathematic equation. An advantage of this method is that data are represented in a compact fashion and in a form which is convenient for differentiation, integration, or interpolation. Frequently the form of the relationship between the dependent and independent variable is known, and it is desired to determine the values of the coefficients in the equation, since these coefficients correspond to physical quantities.

Usually, a known function is given to descript the relationship between the dependent and independent variables. However, in some cases the form of the relationship between the dependent and independent variables is unknown and must be determined. This may be done by plotting the data and comparing the shape with that for known functions. Frequently the functional relationship is such that a straight line graph may be obtained by changing the variables. When the data or some function of the data can be plotted as a straight line, the constant can be determined simply from the slope and intercept. That is, the constants a and b in the linear equation $y=a+bx$. When a straight line is not obtained, it is best to use a power series of the type: $y=a+bx+cx^2+dx^3+\cdots$, with as many empirical constants as necessary to represent the data to within the experimental uncertainty.

When the data follow a linear relationship, there are three procedures for determining the slope and the intercept of the equation that fits the data. These three procedures are the method of graph, the method of average, and the method of least squares, among which the method of least squares is the most tedious but gives the best value for the slope and intercept. If a computer program for least method is designed, not only constants in the linear relationship can be obtained rapidly and precisely, but also a relative coefficient can be given.

第二部分 Part Two 实 验 Experiments

实验一　燃烧热的测定

（一）实验目的

1. 掌握氧弹式量热计使用方法及测量物质燃烧焓的技术。
2. 通过测定萘或蔗糖的燃烧热，掌握恒压燃烧热与恒容燃烧热的区别与联系。
3. 学会应用雷诺图解法校正温度变化值。
4. 了解燃烧热及测定的药学应用。

（二）实验原理

1. 燃烧热测定　燃烧热是指 1mol 有机物完全燃烧时所放出的热量。所谓完全燃烧是指 C 变为 $CO_2(g)$，H 变为 $H_2O(l)$，N 变为 $N_2(g)$，S 变为 $SO_2(g)$，Cl 变为 HCl(aq)，金属变为游离状态等。

燃烧热可在恒容或恒压条件下测定。由热力学第一定律可知，不做非体积功的情况下，恒容燃烧热 $Q_V=\Delta_r U$，恒压燃烧热 $Q_p=\Delta_r H$。用氧弹式量热计测得的是恒容燃烧热 Q_V，而通常化学反应的热效应是用恒压热效应 Q_p 来表示的。若把与反应有关的气体作为理想气体处理，则 Q_p 与 Q_V 之间可通过以下关系换算，即

$$Q_p=Q_V+\Delta nRT \qquad 式(2\text{-}1\text{-}1)$$

式(2-1-1)中 Δn 为反应前后产物和反应物中气体的物质的量之和的差值；R 为气体常数；T 为反应的绝对温度。

使一定量的样品在氧弹中完全燃烧，假设系统和环境之间没有热交换，则系统放出的全部热量可使量热计的温度由 T_1 升高到 T_2，通过测定燃烧前、后量热计温度的变化，就可以计算出样品的燃烧热。其关系式如下：

$$\frac{m}{M}Q_{m,V}+m_{点火丝}Q_{点火丝}+m_{棉线}Q_{棉线}+C(T_2-T_1)=0 \qquad 式(2\text{-}1\text{-}2)$$

式(2-1-2)中 m 和 M 为样品的质量和摩尔质量；$Q_{m,V}$ 为样品的摩尔恒容燃烧热；$m_{点火丝}$ 和 $m_{棉线}$ 为点火丝和棉线的质量；$Q_{点火丝}$ 和 $Q_{棉线}$ 分别为单位质量的点火丝和棉线的恒容燃烧焓(J/g)，$Q_{点火丝}=-3\ 243J/g$，$Q_{棉线}=-16\ 736J/g$；C 为量热计的热容。

由式(2-1-2)可知，若已知量热计的热容，测得量热计温度的变化值，就可求出样品(萘或蔗糖)的摩尔恒容燃烧热，并由式(2-1-1)可计算出摩尔恒压燃烧热。量热计的热容可用标准物质来标定，本实验选用的标准物质为苯甲酸，其摩尔恒容燃烧热 $Q_{m,V}(298K)=$

-3 227kJ/mol。

2. 量热计和氧弹的结构 图 2-1-1 是实验所用氧弹式量热计的装置示意图。量热计主要由精密温度计、水套(外筒)、盛水桶(内筒)、自密封式氧弹、搅拌器、电机等部件组成。为了减少与周围环境的热交换,量热计在设计上采取了使其内外套壳内壁高度抛光、梨形截面盛水桶以减少内筒与外筒间的热辐射作用等措施。搅拌器可使样品燃烧放出的热量尽快在量热计中均匀散布,通过测量精度为 0.001℃ 的精密温度计可准确测出每一时刻系统温度的变化。

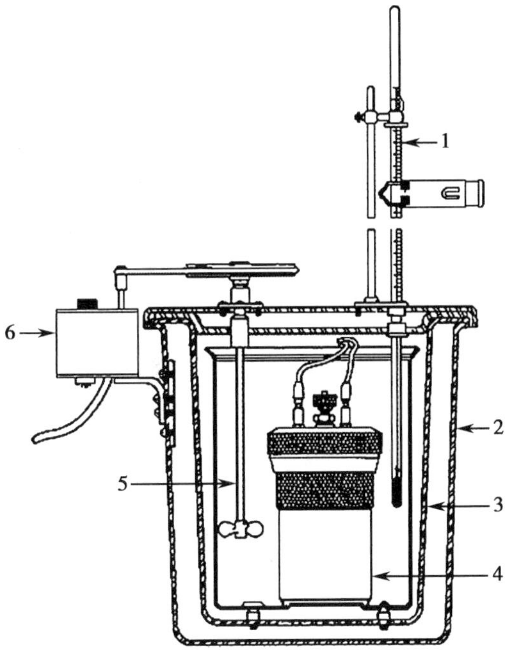

1. 精密温度计(precise thermometer);2. 水套(water pocket);3. 盛水桶(water bucket);4. 氧弹(oxygen bomb);5. 搅拌器(stirrer);6. 电动机(stirring motor)

图 2-1-1 氧弹式量热计装置示意图

Fig. 2-1-1 Sketch of the Oxygen Bomb Calorimeter

氧弹是用高强度耐腐蚀的不锈钢制成的,见图 2-1-2 所示。主要部件有厚壁弹体和弹盖,在弹盖上装有用来充入氧气的进气孔、排气孔和电极终端,电极直通弹体内部,同时作为燃烧皿的支架。氧弹采用自动密封橡胶垫圈,当氧弹内充氧到一定压力时,橡胶垫圈因受压而与弹体和弹盖密接,造成两者间的气密性,且氧弹筒内外压力差越大,密封性能越好。

(三) 仪器和试剂

1. 仪器 氧弹量热计全套,氧气钢瓶,减压阀,压片机,万用电表,温度计(0~50℃),分析天平,1 000ml 量筒,点火丝,棉线。

2. 试剂 苯甲酸(分析纯),萘(分析纯)或蔗糖(分析纯)。

(四) 实验步骤

1. 量热计热容的测定

(1) 样品压片:取约 0.8g 苯甲酸,将钢模底板装入模子中,从上面倒入已称好的样品,缓慢旋紧压片机的螺杆,直到将样品压成片状为止。抽取模底的托板,再继续旋动螺杆,使模底的样品脱落于称量纸上。将药片表面的碎屑轻轻敲去,在分析天平上准确称量后得到样品质量 m。

(2) 装置氧弹:打开氧弹,将氧弹内壁擦净。将称重好的点火丝两端分别固定在氧弹连接环的卡槽中,连接已称重的棉线与点火丝,并使棉线的两端搭在燃烧皿中,再将准确称量的药片压在棉线上。小心盖上并旋紧氧弹盖,用万用电表检查两电极是否通路。若通路,则可以充氧气。

(3) 充氧气:图 2-1-3 为氧弹充气示意图。充氧时,将氧气表头的导管与氧弹的进气口连接,打开氧气钢瓶的主阀 1,观察高压表的指示是否符合要求(至少在 4MPa),慢慢旋紧(即打开)减压阀 2 使低压表指针指向 1.5~2.0MPa,即氧弹中充有 1.5~2.0MPa 的氧气。放松(即关闭)减压阀,取下氧弹,关闭氧气钢瓶上端阀门 1,打开减压阀放掉管道内及氧气表

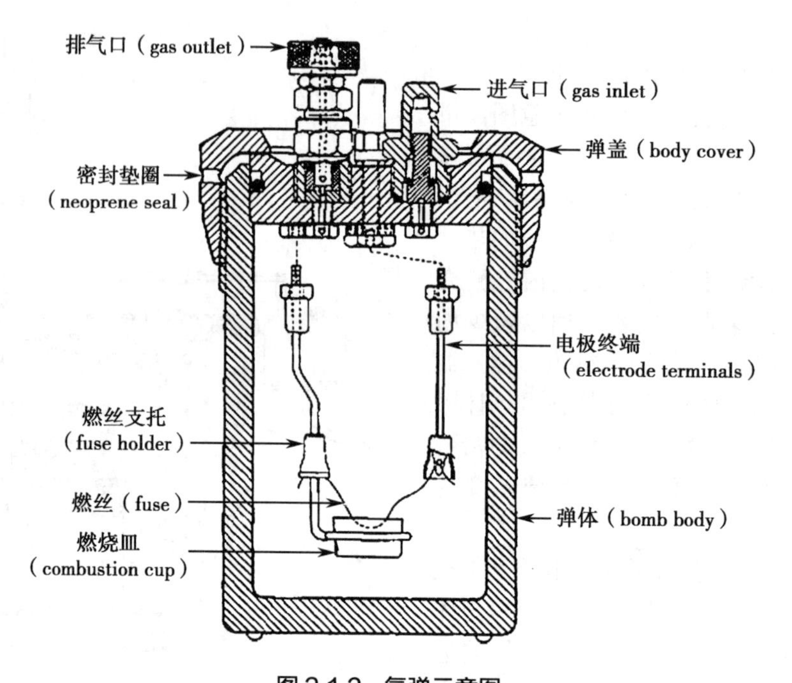

图 2-1-2 氧弹示意图
Fig. 2-1-2 Sketch of Oxygen Bomb

内的余气。

（4）测量温度：用万用电表再次检查氧弹的两极间是否通路。若不通，则需放掉其中的氧气，旋开氧弹盖，重新用点火丝连接两电极，直至连好。用量筒准确量取 2 500ml 自来水，倒入盛水桶内，调节水温比室温低 0.5~1.0℃，将氧弹放入水中，注意水面应没过氧弹，如有气泡逸出，说明氧弹漏气，应寻找原因排除。

接上点火导线，盖上保温盖，将测温传感器插入水中，打开电源和搅拌开关，仪器开始显示水温，当水温均匀上升后，每分钟记录一次温度，记下第 10 个数据的同时按“点火”键（按键时间不超过 5 秒），以后每隔半分钟读取并记下温度。当温度上升不明显时改为一分钟记录一次温度值，再记录 10 个数据即可停止实验。

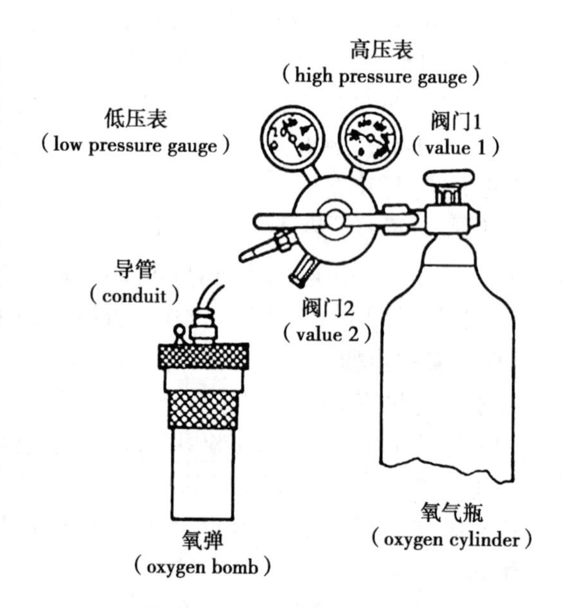

图 2-1-3 氧弹充气示意图
Fig. 2-1-3 Sketch of Oxygen Gas Filling

停止搅拌，切断所有电源，拿出传感器，取出氧弹，用放气阀放掉氧弹内多余气体，开启氧弹，检查样品是否燃烧完全（若留有许多黑色残渣表示燃烧不完全，实验失败，需重复实验），测量剩余点火丝的质量，用酒精棉将氧弹内部擦拭干净。

2. 萘的燃烧热测定　称取约 0.6g 萘，重复上述操作。

3. 蔗糖的燃烧热测定　称取 1.5g 蔗糖，重复上述操作。

注意:2 和 3 选做其一。

4. 清理仪器,整理实验台。

(五)数据记录与处理

1. 列表记录实验数据。

2. 作图法求出苯甲酸燃烧引起量热系统温度的变化值,计算量热系统的热容。

3. 按作图法求出萘或蔗糖燃烧引起的量热系统温度的变化值,并计算萘或蔗糖的恒容燃烧热。

4. 根据式(2-1-1),由萘或蔗糖的恒容燃烧热(Q_V)计算恒压燃烧热(Q_p)。

5. 由物理化学数据手册查出萘或蔗糖的恒压燃烧热,计算实验误差,分析误差来源。

(六)思考题

1. 说明恒容燃烧热(Q_V)和恒压燃烧热(Q_p)的区别和联系。

2. 使用氧气钢瓶和减压阀时要注意什么问题?

3. 如何用萘的燃烧热数据来计算萘的生成热?

(七)药学应用

在热化学中一般应用量热计来测定物质的热力学性质。燃烧热是热化学中重要的基本数据,通过这些数据可间接求算许多化合物的生成热、化学反应的热效应和键能等。

此外,依据同素异构体分子燃烧焓的差别,进而可推断其稳定性。通过对不同厂家的药物制剂(如阿奇霉素片)、不同地区的中药材或药用植物(如艾叶)等的燃烧热的测定,可确定药物的稳定性,为药物的质量评价提供理论依据。

(八)雷诺图解法

为准确求得样品燃烧前后量热计温度的变化值(ΔT),须以温度对时间作图加以校正。由于量热计不能做到完全绝热,会有热传导、蒸汽、对流、辐射等引起的热交换以及搅拌所引入的搅拌热等,它们对温度测量的影响必须扣除。实验中采用雷诺图解法对 ΔT 进行温度校正,其方法如下:

(1)将观察到的温度对时间作图(图 2-1-4),PA 表示热效应开始前的温度变化,BQ 表示热效应终了时的温度变化。

(2)作 AB 间的垂线 BR,通过 BR 的中点 C 作平行于横轴的直线交曲线 AB 于 D 点。

(3)过 D 点作一垂线分别交 BQ 和 PA 的延长线于 F 点和 E 点,EF 即为所求的 ΔT。

有时量热计的绝热良好,热量散失少,但却由于搅拌器的功率较大,使得量热计在热效应终了时体系温度随时间缓慢增加,这种情况下 ΔT 仍然可以按照此法进行校正。

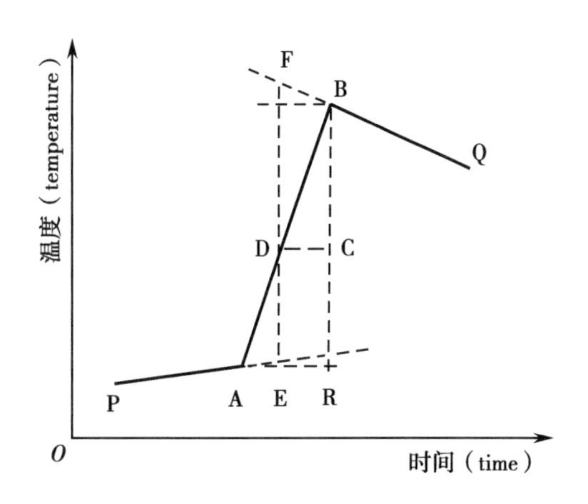

图 2-1-4 雷诺温度校正图

Fig. 2-1-4 Reynolds temperature Correction Curve

Experiment 1 Determination of Heats of Combustion

1. Objective

1.1. To study the operation method of a oxygen bomb calorimeter, to learn the experimental technique of calorimetry.

1.2. To measure the heats of combustion of naphthalene or sucrose, and to study the difference and relationship between heats of combustion at constant volume and constant pressure.

1.3. To calibrate the temperature variation by Reynolds correction curve.

1.4. To learn the pharmaceutical application of heat of combustion and its determination.

2. Principle

2.1. Combustion determination

The heat of combustion for a substance is the heat released from a reaction in which one mole of a compound containg C, H, N, S, Cl or metal etc. undergoes complete combustion to form specified combustion products, such as $CO_2(g)$, $H_2O(l)$, $N_2(g)$, $SO_2(g)$, HCl(aq), or free state metal etc., correspondly.

Depending on the experimental conditions for combustion, the heat released, Q, equals to either Q_V or Q_p. According to the first law of thermodynamics, Q_V equals to the thermodynamic energy change, $\Delta_r U$, at constant volume; Q_p equals to the enthalpy change, $\Delta_r H$, at constant pressure. Using an adiabatic bomb calorimeter (Fig. 2-1-1), Q_V is measured because the reaction is carried out in a closed bomb with constant volume. Then the Q_p can be calculated by the following equation:

$$Q_p = Q_V + \Delta nRT \tag{2-1-1}$$

where Δn is the difference of number of moles between the gaseous products and the gaseous reactants, R is the molar gas constant, and T is the reaction temperature (K). Note that equation (2-1-1) can only be obtained under the condition that all gaseous reactants and products can be taken as ideal gases.

Assuming there is no heat transfer between the system and surroundings, the heat released will lead to the temperature rise of the system from T_1 to T_2 when completely burning m gram of the sample in a bomb. Therefore, the heat of combustion can be calculated by measuring the temperature change of the calorimeter before and after combustion. The calculation equation is as follows:

$$\frac{m}{M}Q_{m,V} + m_{fuse}Q_{fuse} + m_{cotton}Q_{cotton} + C(T_2 - T_1) = 0 \tag{2-1-2}$$

where m and M are mass and molar mass of the sample respectively; $Q_{m,V}$ is the molar heat of combustion of the sample at constant volume; m_{fuse} and m_{cotton} are mass of fuse and cotton respectively; Q_{fuse} and Q_{cotton} are heats of combustion of fuse and cotton at constant volume respectively, using −3 243J/g for fuse and −16 736J/g for cotton; C is the heat capacity of the calorimeter which can be calibrated using a standard sample with known heat of combustion.

Once the calorimeter has been calibrated, the heat of combustion for an unknown sample, naphthalene or sucrose, will be calculated using equation(2-1-2). In this experiment, benzoic acid combustion with $Q_{m,V}(298K) = -3\ 227kJ/mol$ is used for such purpose.

2.2. Calorimeter and bomb

The main parts of the bomb calorimeter are shown schematically in Fig. 2-1-1. The calorimeter consists of a metal reaction chamber(bomb) that is immersed in a water bath with a known volume of water, which prevents the effect of temperature variation of surroundings on the system. The temperature rise accompanying the combustion is read from a precise thermometer with digital output whose probe is immersed in the water. A stirrer effects an even distribution of the temperature. The water bucket in turn is surrounded by an insulating air space, which prevents, as far as possible, heat leakage to the surroundings. A precision thermometer with a measurement accuracy of 0.001℃ can accurately measure the change in system temperature at each moment.

Oxygen bomb is made of stainless metal as shown in Fig. 2-1-2. The main parts are cylinder with thick wall, cover and screw. There is gas inlet for oxygen filling, gas outlet and electrode which is through the bomb to be the support of combustion cup. Inside this bomb, the sample held in the cup is ignited by passing electrical current through a fuse wire and cotton line. In the combustion process, fuse wire and cotton line are also consumed. The interior of the reaction chamber is pressurized with oxygen to ensure efficient combustion of the material of interest.

3. Apparatus and Chemicals

3.1. Apparatus

Bomb calorimeter system, oxygen gas cylinder, pressure reducing valve, tablet machine, multi-meter, precise thermometer with digital output, thermometer(0~50℃), analytical balance, 1 000ml measuring cylinder, fuse wire, cotton line.

3.2. Chemicals

Benzoic acid(AR), naphthalene(AR) or sucrose(AR).

4. Procedures

4.1. Measurement of heat capacity of the bomb calorimeter

4.1.1. Preparation of tablet: Weigh out approximately 0.8g of benzoic acid. Make the benzoic acid into tablet by means of the tablet machine, get rid of the powder on the surface gently, and then weigh the tablet precisely using an analytical balance.

4.1.2. Assembly of the bomb calorimeter: Open the bomb, clean and dry all parts of it. Weigh accurately the fuse wire and the cotton line. Set the bomb head in the support stand and attach the fuse wire. Make sure that the wire does not touch the combustion cup. Wrap the cotton line around the fuse wire and put two ends of the cotton line in the combustion cup. Place the sample tablet right on the cotton line in the cup with the pincette. Cover the bomb carefully and tightly. Check the circuit with a multi-meter. If the circuit is conductible, fill the bomb with oxygen gas.

4.1.3. Oxygen gas filling: Fig. 2-1-3 is the schematic of oxygen gas filling. Slip on the oxygen tank connection hose to the pin on the head assembly. Open the main valve on the oxygen tank slowly and make sure it shows pressure(>4MPa). Open the regulator valve slowly and fill the bomb with oxygen gas until the bomb pressure rises to 1.5~2.0MPa. Close the control valve and then the tank valve. Open the quick-release valve to quickly remove the oxygen tank connection, and release the filling hose.

4.1.4. Measurement of temperature: Check the circuit again with the multi-meter to make sure that the circuit is conductible. Release the oxygen in the bomb and reconnect the fuse wire in case that the circuit is cut off. Fill the water bucket with 2 500ml of water, adjust the temperature to about 0.5 ~ 1.0℃ below the room temperature. Immerse the bomb into the water. If gas bubbles escape from the bomb, the assembly ring may require tightening, or the gaskets may need to be replaced.

Connect the fuse wire, set the cover on the jacket and insert the probe of the precise thermometer into the calorimeter so that the end of sensor touches the water. Switch on the power, start the stirrer to be sure that it runs freely. Record the temperature of the bucket every minute precisely (with a precision up to 0.001℃) when it starts to increase steadily. After 10min, fire the bomb by pressing the ignition button and holding it down for no more than 5s because it may damage the ignition unit or result in undue heating by passage of current through the water. Then, record the temperature every 30s till the rate of temperature change becomes small enough, and then record the temperature at 1min intervals again for a further 10min.

After the last temperature reading, turn off all the electrical connections, take off the probe, and remove the bomb. Relieve the pressure by opening the valve, and open the bomb to check the completion of combustion. Do the experiment again if any evidence of incomplete combustion is observed. Weigh any unburned fuse wire still attached to the electrodes and possible pieces of molten wire. When analyzing the results, subtract this weight from the total fuse wire burned. Clean and dry the bomb and calorimeter after each experiment.

4.2. Measurement of the heat of combustion of naphthalene

Weigh out approximately 0.6g of naphthalene and repeat the procedure described above.

4.3. Measurement of the heat of combustion of sucrose

Weigh out approximately 1.5g of sucrose and repeat the procedure described above.

Attention: choose 4.2 or 4.3 to perform the experiment.

4.4. Clean up the instrument and tidy up the test bench.

5. Data recording and processing

5.1. Draw tables to record the data.

5.2. Plot the temperature-time curve for combustion of benzoic acid, make the Reynolds correction, and calculate the heat capacity of the calorimeter from ΔT.

5.3. Calculate the heats of combustion respectively at constant volume for naphthalene or sucrose.

5.4. Calculate the heats of combustion respectively at constant volume (Q_p) for naphthalene or sucrose using equation (2-1-1) according to their heats of combustion respectively at constant volume (Q_V).

5.5. Find the heat of combustion for naphthalene or sucrose from the handbook of physical chemistry data and calculate the experimental error. Analyze the source of inaccuracy.

6. Questions

6.1. Explain the difference and relationship between heat of combustion at constant volume,

Q_V, and heat of combustion at constant pressure, Q_p.

6.2. How careful must you be in operating the oxygen gas cylinder and reducing pressure valve?

6.3. How to calculate the formation heats of naphthalene and sucrose based on the heats of combustion?

7. Pharmaceutical applications

A calorimeter is commonly used to determine the thermodynamic properties of substances. Heats of combustion are important fundamental parameters of substances since these values can be used to speculate the heat of formation, heat effect of chemical reactions and bond energy.

In addition, the measurements of heat of combustion of isomers are also used to evaluate their stability. By measuring heat of combustion of medicine preparations from different manufacturers (such as Azithromycin tablets) and Chinese medicinal materials or medicinal plants from different regions (such as Artemisia argyi), the stability and quality evaluation of drugs can be provided.

8. Reynolds correction curve

To obtain the ΔT precisely, a temperature-time curve is plotted. Because the calorimeter is not absolutely adiabatic, the heat transfer that is caused by heat conducting, evaporating, convecting and emitting, between system and surroundings cannot be avoided. In this experiment, Reynolds temperature correction curve is used to calibrate the ΔT. The detailed procedure is as follows.

(1) Plot the recorded value of the system temperature versus time, shown in Fig. 2-1-4. PA is the temperature variation before the beginning of heat effect, and BQ is the temperature variation after the ending of heat effect.

(2) Draw a vertical line of point B to get line BR. Draw a horizontal line through the middle point C of line BR to point D on line AB.

(3) Then draw a vertical line of point D, points E and F can be obtained by extending BQ and PA lines to this vertical line. Therefore, the temperature difference between point E and F is considered as ΔT for calculation.

Sometimes the calorimeter has good thermal insulation and less heat loss, but due to the large power of the agitator, the temperature of the system slowly increases with time at the end of the heat effect, in which case ΔT can still be corrected according to this method.

实验二 溶解热曲线的测定

(一) 实验目的

1. 掌握电热补偿法测定热效应的原理和方法。

2. 应用电热补偿法测定硝酸钾的积分溶解热曲线，并求一定浓度时的积分稀释热、微分溶解热和微分稀释热。

3. 了解溶解热测定的药学应用。

(二) 实验原理

1. 溶解热概念 等温等压下，一定量的物质溶于一定量的溶剂中所产生的热效应称为该物质的溶解热。溶质溶解于溶剂的过程由溶质晶格破坏、电离的吸热过程和溶质溶剂化的放热过程组成，总的热效应取决于两者之和，溶解过程吸热和放热都有可能。在一定温度和压力下，热效应的大小与溶质和溶剂的相对量有关，例如硝酸钾溶解在水中的热效应（吸热）随溶剂水的量增加而增加。

在标准压力下，1mol 溶质溶于 n_0 mol 溶剂中的热效应，称为浓度 1∶n_0 时的积分溶解热 $\Delta H_m^\ominus$。当 n_0 取不同值时，以 $\Delta H_m^\ominus$ 对 n_0 作图，所得曲线称为积分溶解热曲线，如图 2-2-1 所示。本实验测定的是硝酸钾的积分溶解热曲线。当 1mol 溶质从浓度 1 稀释到浓度 2 时的热效应，称为积分稀释热。可由两个浓度下的积分溶解热之差求得，即

$$\Delta H_m^\ominus(\text{稀释}) = \Delta H_m^\ominus(2) - \Delta H_m^\ominus(1) \qquad \text{式(2-2-1)}$$

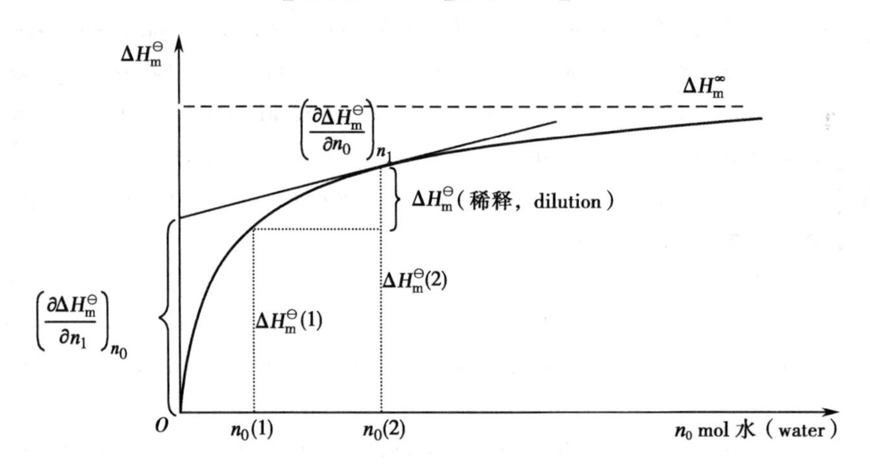

图 2-2-1　积分溶解热曲线

Fig. 2-2-1　Curve of Integral Heat of Solution

由于热效应的大小与溶质的量 n_1 和溶剂的量 n_0 有关，根据偏摩尔量的集合公式

$$\Delta H_m^\ominus = \left(\frac{\partial \Delta H_m^\ominus}{\partial n_0}\right)_{T,p,n_1} n_0 + \left(\frac{\partial \Delta H_m^\ominus}{\partial n_1}\right)_{T,p,n_0} n_1 \qquad \text{式(2-2-2)}$$

式中 $\left(\frac{\partial \Delta H_m^\ominus}{\partial n_0}\right)_{T,p,n_1}$ 为微分稀释热，其物理意义是在一定温度、压力且溶质量恒定时，在足量的溶液中，溶剂的量变化 1mol 时引起的热效应；$\left(\frac{\partial \Delta H_m^\ominus}{\partial n_1}\right)_{T,p,n_0}$ 为微分溶解热，其物理意义是在一定温度、压力且溶剂量恒定时，在足量的溶液中，溶质的量变化 1mol 时引起的热效应。对于溶质的量恒定为 1mol 的系统，即 $n_1=1$ 时，式(2-2-2)简化为

$$\Delta H_m^\ominus = \left(\frac{\partial \Delta H_m^\ominus}{\partial n_0}\right)_{T,p,n_1} n_0 + \left(\frac{\partial \Delta H_m^\ominus}{\partial n_1}\right)_{T,p,n_0} \qquad \text{式(2-2-3)}$$

因此，在 $\Delta H_m^\ominus \sim n_0$ 图中，在某一浓度处，例如图 2-2-1 中的 $n_0(2)$ 处作一条切线，切线的斜率就是该浓度时的微分稀释热 $\left(\frac{\partial \Delta H_m^\ominus}{\partial n_0}\right)_{T,p,n_1}$，切线的截距就是该浓度时的微分溶解热 $\left(\frac{\partial \Delta H_m^\ominus}{\partial n_1}\right)_{T,p,n_0}$。由此，在 $\Delta H_m^\ominus \sim n_0$ 图中可以同时得到积分溶解热、积分稀释热、微分溶解热和微分稀释热的数据。

2. 电热补偿法原理　硝酸钾溶解于水的过程是吸热过程，反应热可以用电热补偿法来

进行测定。其基本做法是，在反应前确定系统的温度，在反应中，给予系统电加热，直到反应结束后，系统的温度恢复到起始状态，计算电热量即为反应热。电热量 Q 可以由电流强度 I、电压 V 和通电时间 t 算得，即

$$Q=IVt \qquad 式(2\text{-}2\text{-}4)$$

则积分溶解热为

$$\Delta H_{\mathrm{m}}^{\ominus}=\frac{Q}{n} \qquad 式(2\text{-}2\text{-}5)$$

式中 n 为已溶解硝酸钾的物质的量。

3. 已知标准物法　五水合硫代硫酸钠的积分溶解热可用标准物法测得。用一已知积分溶解热的标准物质（本实验采用氯化钾）标定出量热计的热容量 C，然后根据待测样品溶解前后量热系统温度的变化，可求得待测物质的积分溶解热，即为

$$\Delta H_{\mathrm{m},2}^{\ominus}=\frac{CM_2\Delta T_2}{W_2}=\frac{W_1\Delta H_{\mathrm{m},1}^{\ominus}}{M_1\Delta T_1}\times\frac{M_2\Delta T_2}{W_2} \qquad 式(2\text{-}2\text{-}6)$$

式中 C 为量热系统的热容量；W_1、M_1 为标准物质的质量和摩尔质量；W_2、M_2 分别为待测物质的质量和摩尔质量；$\Delta H_{\mathrm{m},1}^{\ominus}$ 为标准物质在某溶液温度及浓度下的积分溶解热，此值可由手册查到；$\Delta H_{\mathrm{m},2}^{\ominus}$ 为待测物质的积分溶解热；ΔT_1 为标准物质溶解前后量热系统的温度变化值；ΔT_2 为待测物质溶解前后量热系统的温度变化值。

4. 溶解热曲线的经验方程　硝酸钾的溶解热曲线较好地符合以下经验方程，即

$$\Delta H_{\mathrm{m}}^{\ominus}=\frac{\Delta H_{\mathrm{m}}^{\infty}n_0}{b+n_0} \qquad 式(2\text{-}2\text{-}7)$$

式中 $\Delta H_{\mathrm{m}}^{\infty}$ 为溶剂 $n_0\to\infty$ 时的溶解热，即极限溶解热；b 为经验常数。以上两个参数可以通过下式线性拟合得到

$$\Delta H_{\mathrm{m}}^{\ominus}=\Delta H_{\mathrm{m}}^{\infty}-\frac{\Delta H_{\mathrm{m}}^{\ominus}}{n_0}b \qquad 式(2\text{-}2\text{-}8)$$

即将 $\Delta H_{\mathrm{m}}^{\ominus}\sim\frac{\Delta H_{\mathrm{m}}^{\ominus}}{n_0}$ 线性回归，由截距得到极限溶解热 $\Delta H_{\mathrm{m}}^{\infty}$，由斜率的负值得到 b。由式(2-2-8)还可以计算微分稀释热和微分溶解热，即

$$\left(\frac{\partial\Delta H_{\mathrm{m}}^{\ominus}}{\partial n_0}\right)_{T,p,n_1}=\left[\partial\left(\frac{\Delta H_{\mathrm{m}}^{\infty}n_0}{b+n_0}\right)\Big/\partial n_0\right]_{T,p,n_1}=\frac{\Delta H_{\mathrm{m}}^{\infty}b}{(b+n_0)^2} \qquad 式(2\text{-}2\text{-}9)$$

$$\left(\frac{\partial\Delta H_{\mathrm{m}}^{\ominus}}{\partial n_1}\right)_{T,p,n_0}=\Delta H_{\mathrm{m}}^{\ominus}-\left(\frac{\partial\Delta H_{\mathrm{m}}^{\ominus}}{\partial n_0}\right)_{T,p,n_1}n_0=\frac{\Delta H_{\mathrm{m}}^{\infty}n_0^2}{(b+n_0)^2} \qquad 式(2\text{-}2\text{-}10)$$

（三）仪器和试剂

1. 仪器　500ml 量筒，直流电源，数字电压表（精度 0.001V），数字电流表（精度 0.001A），量热计（包括保温杯和加热器），磁力搅拌器，数字贝克曼温度计，计时表，分析天平，台式天平。

2. 试剂　硝酸钾（分析纯），五水合硫代硫酸钠（分析纯），氯化钾（分析纯）。

（四）实验步骤

1 和 2 选做其一。

1. 电热补偿法测定硝酸钾积分溶解热

(1) 硝酸钾研细后存放在干燥器中。

(2) 用分析天平称取 8 份硝酸钾，质量分别为 2.5g、1.5g、2.5g、3g、3.5g、4g、4g 和 4.5g，均

存放在干燥器中备用。

（3）将干燥的保温杯置于台式天平上，加入蒸馏水，精确称量至 216.2g（12mol），同时记录水温，作为实验温度。

（4）按图 2-2-2 搭好装置，并连接好导线（注意加热器一头导线先断开），开启磁力搅拌器，调节转速。开启电源，接上加热器，调整功率约为 2.5 伏安（电压约 5V，电流约 0.5A），准确记录电压值和电流值。当数字贝克曼温度计读数上升 0.5℃时，记作标记温度，并按下秒表开始计时。

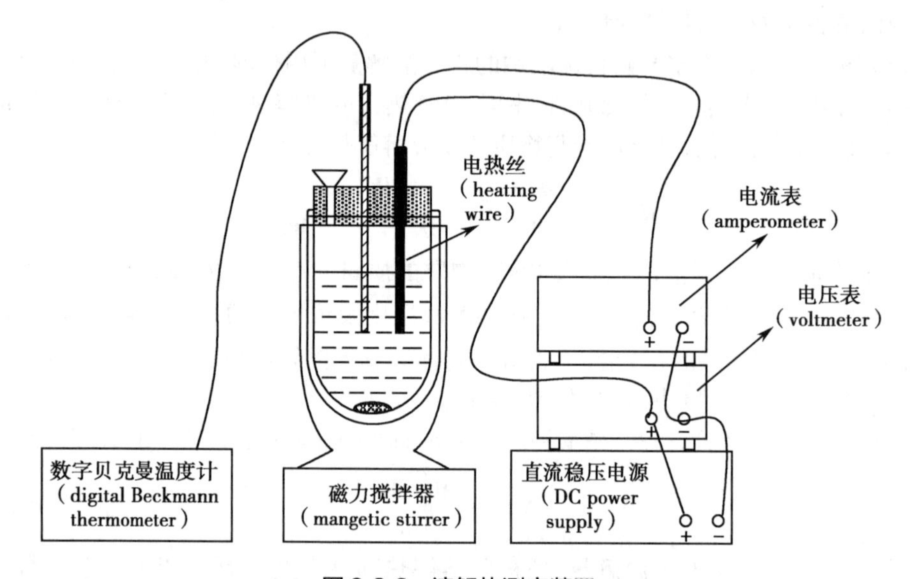

图 2-2-2 溶解热测定装置

Fig. 2-2-2 Sketch of the Calorimeter

（5）慢慢加入第一份样品（温度迅速降低），当样品全部加入后，取出漏斗，加塞封口。待温度恢复到标记温度时，记下加热时间（不要按停秒表）。接着通过漏斗加入第二份样品。

（6）依次重复步骤 5，直到测完 8 份样品。

（7）清洗仪器，整理实验台。

2. 标准物质法测定五水合硫代硫酸钠或硝酸钾积分溶解热

（1）测定量热系统的热容：用溶解热已知的氯化钾作为标准物质来标定量热系统的热容，不同温度下 1mol 氯化钾溶于 200ml 水中的积分溶解热数据见附录表 4-6。实验装置如图 2-2-2 所示，但不含加热装置。

用 500ml 量筒量取 360ml 蒸馏水于杜瓦瓶中，盖严瓶塞，缓慢均匀搅拌，使蒸馏水与量热系统的温度达到平衡。每分钟读取温度一次，当连续 5 分钟温度读数不变时可认为已达到平衡，此温度即为 $T_{始}$。

将预先称好的氯化钾（7.5±0.01）g 全部迅速倒入杜瓦瓶中，塞好瓶塞，缓慢均匀地搅拌。由于氯化钾溶解为吸热过程，溶解时温度下降，每分钟读取温度一次，直至 5 分钟内温度不变，即为 $T_{终}$。再用普通温度计测出量热计的温度。倒出量热计中的液体，洗净并晾干。

（2）硝酸钾积分溶解热的测定：用硝酸钾代替氯化钾重复上述操作，测出 $T_{始}$ 和 $T_{终}$，硝酸钾的用量按 1mol KNO_3 ∶ 400mol 水计算。其用量约为 5.1g，蒸馏水仍为 360ml。

（3）五水合硫代硫酸钠积分溶解热的测定：同（2）。测定 1mol $Na_2S_2O_3 \cdot 5H_2O$ ∶ 400mol

水的积分溶解热，水仍为 360ml，五水合硫代硫酸钠用量约为 12.4g。

注意：(2)和(3)选做其一。

（五）实验步骤

1. 按表 2-2-1 记录实验数据，其中 $n_0=\dfrac{12}{n}$，$Q=IVt$，$\Delta H_{\mathrm{m}}^{\ominus}=\dfrac{Q}{n}$。

表 2-2-1 硝酸钾溶解热曲线的测定

Table 2-2-1 Determination of Heat of Solution of Potassium Nitrate

水的质量(mass of water) m：________ g；电流(current) I：________ A

No.	KNO_3			n_0/mol	t/s	Q/J	$\Delta H_{\mathrm{m}}^{\ominus}$/(J/mol)	$\Delta H_{\mathrm{m}}^{\ominus}/n_0$
	m/g	$M_{累积\ (accumulated)}$/g	n/mol					
1								
2								
…								
7								
8								

2. 作 $\Delta H_{\mathrm{m}}^{\ominus}\sim n_0$ 图。

3. 作 $\Delta H_{\mathrm{m}}^{\ominus}\sim\dfrac{\Delta H_{\mathrm{m}}^{\ominus}}{n_0}$图，求参数 $\Delta H_{\mathrm{m}}^{\infty}$ 和 b。

4. 计算 $n_0=100$mol 和 $n_0=200$mol 时的积分溶解热，以及 n_0 从 100mol 改变至 200mol 时的积分稀释热。

5. 计算 $n_0=200$mol 时的微分溶解热和微分稀释热。

6. 按公式(2-2-6)计算硝酸钾或五水合硫代硫酸钠的积分溶解热。

（六）思考题

1. 若溶解为放热过程，能否用电热补偿法测定其溶解热？

2. 电热补偿法能否用来测定液体的比热容、水化热、生成热和液体混合热？

3. 为什么开始时系统的温度要高出环境温度 0.5℃？

4. 测定 KNO_3 溶解热实验中，应怎样正确操作以减少误差？哪些情况会影响实验结果？

（七）药学应用

固体药物制剂口服给药后，药物的吸收取决于药物从制剂中的溶出或释放、药物在生理条件下的溶解过程。溶解时吸热或放热对药物溶出有很大影响。有研究表明，溶解热与药物的溶出速率、溶解度有关。因此溶解热测定可用于药物质量的控制。溶解热与固体晶格能有关，反映了固体的结晶性，所以溶解热也是鉴别固体药物多晶型的一种方式。

（八）传统贝克曼温度计的调节和使用

在科学研究和物理化学实验中，常常需要对系统的温度差进行精确的测量，然而普通温度计不能达到此精确度，需用贝克曼温度计进行测量。

玻璃水银贝克曼温度计是一种用来精密测量系统温度变化值的水银温度计，其构造如图 2-2-3a 所示。传统的贝克曼温度计长达 40~50cm，温度计上的标度通常只有 5℃，每一度分成100等份,因此贝克曼温度计可以直接读出0.01℃,若用放大镜观察,可以估读到0.002℃。贝克曼温度计设计的特点是在温度计的上部有水银储槽,通过调节水银球中的汞量使在所测定的温度范围内温差都能指示出来。

调节方法一:将温度计浸在温度较高的恒温浴中,然后取出,将其倒置,使毛细管中的水银与贮管中的水银相连。另取一恒温浴,调节温度比最高读数温度高1~2℃,把温度计置于其中,恒温5分钟。取出温度计,垂直握紧,用左手轻敲右小臂,水银柱在弯头处断开。将调节好的贝克曼温度计置于待测系统中,观察其示值是否达到预定位置,若不符要求,需重复操作至达到要求。(图2-2-3b)

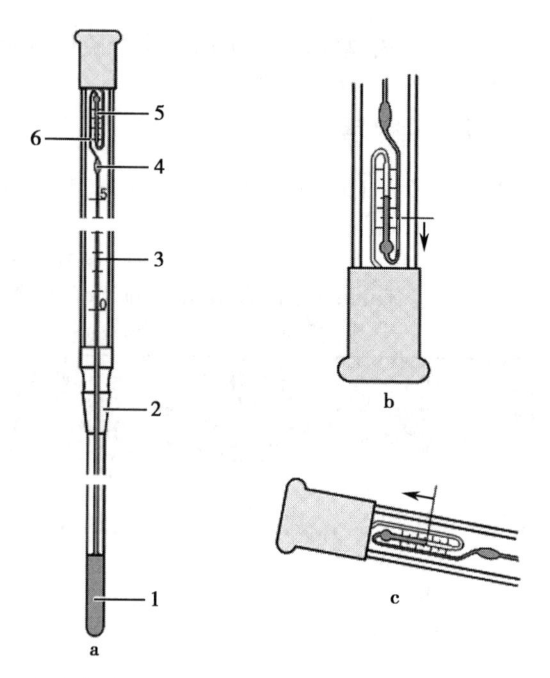

1. 水银球(mercury bulb);2. 磨口(grinding);3. 水银线(mercury thread);4. 溢出泡(overflow bulb);5. 水银贮槽(mercury reservoir);6. 毛细管(capillary)

图2-2-3 贝克曼温度计的构造及调节

Fig. 2-2-3 Construction and Setting of Beckmann Thermometer

调节方法二:直接利用温度计头部的标尺。首先估计最高使用温度。将温度计倒置,使水银球和毛细管中的水银流入毛细管末端,然后慢慢倾斜温度计(图2-2-3c),使贮槽中的水银与毛细管中的水银相连接,如图2-2-3c所示。若估计值高于室温,可利用温水或重力作用让水银流入贮槽,当标尺的水银面到达所需温度时,轻轻敲击,使水银柱在弯头处断开。若估计值低于室温,则将温度计浸于较低恒温浴中,当标尺处的水银面下降至温度估计值时,用同法断开水银柱。与方法一相同,试验水银量调节是否合适。

注意:贝克曼温度计比一般温度计长,易损坏。它只能被安装实验仪器上,放在温度计盒中,或握在手中,而不得随意摆放。调节时,避免其受剧热或剧冷,或重击。不允许在一次实验中基温选择换挡。

Experiment 2 Determination of Heats of Solution Curve

1. Objective

1.1. To learn the principle and procedures of electrothermic compensation method for measuring heat effect.

1.2. To measure the integral heats of solution of potassium nitrate or sodium thiosulfate pentahydrate, or to calculate integral heat of dilution, differential heat of solution and differential heat of dilution of potassium nitrate at a specified composition.

1.3. Learn about the pharmaceutical applications of dissolution calorimetry.

2. Principle

2.1. Heat of solution

The process of dissolution includes the damage of the crystal lattice accompanied with the ionization of solutes and the interaction between the solvent and the solute molecules or ions, which is endothermic reaction for the former and exothermic reaction for the latter. The heat of solution is the sum of the heat involved in all the processes. Therefore, it can be either positive or negative based on the relative contents of solvents and solutes.

At a standard pressure, the integral heat of solution ($\Delta H_m^\ominus$) is the heat of reaction when 1 mole of solute is dissolved in n_0 mole of solvent to produce a solution of the given concentration. Thus, the plot of $\Delta H_m^\ominus$ against n_0 is called the curve of integral heat of solution (Fig. 2-1-1). In this experiment, the curve illustrating the integral dissolution heat of potassium nitrate is determined. The integral heat of dilution can be calculated based on the difference between the integral heats of solution at two different concentrations as follows

$$\Delta H_m^\ominus(\text{dilution}) = \Delta H_m^\ominus(2) - \Delta H_m^\ominus(1) \qquad (2\text{-}2\text{-}1)$$

The heat effect $\Delta H_m^\ominus$ depends on the number of moles of solute (n_1) and solvent (n_0). It can be determined based on the following equation

$$\Delta H_m^\ominus = \left(\frac{\partial \Delta H_m^\ominus}{\partial n_0}\right)_{T,p,n_1} n_0 + \left(\frac{\partial \Delta H_m^\ominus}{\partial n_1}\right)_{T,p,n_0} n_1 \qquad (2\text{-}2\text{-}2)$$

where $\left(\frac{\partial \Delta H_m^\ominus}{\partial n_0}\right)_{T,p,n_1}$ is the differential heat of dilution. It means the reaction heat when one mole of solvent is added in sufficient solution at certain temperature, pressure and number of moles of solute. $\left(\frac{\partial \Delta H_m^\ominus}{\partial n_1}\right)_{T,p,n_0}$ is the differential heat of solution which means the reaction heat when one mole of solute is added in sufficient solution at certain temperature, pressure and number of moles of solvent. For a system of 1mole solute (i.e., $n_1 = 1$), equation (2-2-2) becomes

$$\Delta H_m^\ominus = \left(\frac{\partial \Delta H_m^\ominus}{\partial n_0}\right)_{T,p,n_1} n_0 + \left(\frac{\partial \Delta H_m^\ominus}{\partial n_1}\right)_{T,p,n_0} \qquad (2\text{-}2\text{-}3)$$

Hence, the slope of the tangent line for the plot of the integral heat of solution at a certain amount of solvent, say $n_0(2)$ shown in Fig. 2-2-1, is the differential heat of dilution at the composition specified. The intercept of the tangent line is the differential heat of solution at the certain concentration. Consequently, integral heat of solution, integral heat of dilution, differential heat of solution and differential heat of dilution can be obtained based on the plot of $\Delta H_m^\ominus$ against n_0.

2.2. Principle of electrothermic method

The process of dissolution of potassium nitrate (KNO_3) in water is an endothermic reaction. To compensate the decrease of temperature, electrothermic method is used to provide enough heat to raise the temperature of the system. The solution is heated by electrically heating wire until the end of the dissolution reaction so that the temperature of the system reverts its initial temperature. The calculation of the electrical heat is the heat of reaction. Electrical heat Q can be calculated from current intensity I, voltage V and power-on time t, that is

$$Q = IVt \tag{2-2-4}$$

Therefore, the integral enthalpy of KNO_3 becomes

$$\Delta H_{\mathrm{m}}^{\ominus} = \frac{Q}{n} \tag{2-2-5}$$

where n is the mole number of KNO_3 dissolved in solution.

2.3. Standard substance method

The integral enthalpy of solution of sodium thiosulfate pentahydrate($Na_2S_2O_3 \cdot 5H_2O$) can be determined using standard substance method. In this experiment, the capacity of the calorimeter can be calibrated using potassium chloride(KCl), a standard substance with known integral heat of solution. The integral heat of solution of $Na_2S_2O_3 \cdot 5H_2O$ can be calculated by the following equation

$$\Delta H_{\mathrm{m},2}^{\ominus} = \frac{CM_2\Delta T_2}{W_2} = \frac{W_1\Delta H_{\mathrm{m},1}^{\ominus}}{M_1\Delta T_1} \times \frac{M_2\Delta T_2}{W_2} \tag{2-2-6}$$

where C is the capacity of the calorimeter, W_1 and M_1 are the mass and molar mass of the standard substance, W_2 and M_2 are mass and molar mass of the sample, $\Delta H_{\mathrm{m},1}^{\ominus}$ is the integral heat of the standard substance, $\Delta H_{\mathrm{m},2}^{\ominus}$ is the integral heat of the sample, ΔT_1 is the temperature change of the system during the dissolution of the standard substance, and ΔT_2 is the temperature change of the system arising from dissolution of the sample, respectively.

2.4. Experiential equation for the plot of dissolution heat

The plot of the dissolution heat for potassium nitrate can be calculated based on the following experiential equation

$$\Delta H_{\mathrm{m}}^{\ominus} = \frac{\Delta H_{\mathrm{m}}^{\infty} n_0}{b+n_0} \tag{2-2-7}$$

where $\Delta H_{\mathrm{m}}^{\infty}$ is the limiting heat of solution at $n_0 \to \infty$, and b is the constant. Equation(2-2-7) can be rearranged to give the linear equation as follows

$$\Delta H_{\mathrm{m}}^{\ominus} = \Delta H_{\mathrm{m}}^{\infty} - \frac{\Delta H_{\mathrm{m}}^{\ominus}}{n_0} b \tag{2-2-8}$$

Thus, a graph of $\Delta H_{\mathrm{m}}^{\ominus}$ against $\Delta H_{\mathrm{m}}^{\ominus}/n_0$ can be plotted, and then $\Delta H_{\mathrm{m}}^{\infty}$ and b can be obtained from the intercept and slope. Equation (2-2-8) can also be used to calculate the differential heat of dilution and differential heat of solution respectively

$$\left(\frac{\partial \Delta H_{\mathrm{m}}^{\ominus}}{\partial n_0}\right)_{T,p,n_1} = \left[\partial\left(\frac{\Delta H_{\mathrm{m}}^{\infty} n_0}{b+n_0}\right)\Big/\partial n_0\right]_{T,p,n_1} = \frac{\Delta H_{\mathrm{m}}^{\infty} b}{(b+n_0)^2} \tag{2-2-9}$$

$$\left(\frac{\partial \Delta H_{\mathrm{m}}^{\ominus}}{\partial n_1}\right)_{T,p,n_0} = \Delta H_{\mathrm{m}}^{\ominus} - \left(\frac{\partial \Delta H_{\mathrm{m}}^{\ominus}}{\partial n_0}\right)_{T,p,n_1} n_0 = \frac{\Delta H_{\mathrm{m}}^{\infty} n_0^2}{(b+n_0)^2} \tag{2-2-10}$$

3. Apparatus and chemicals

3.1. Apparatus

500ml measuring cylinder, DC power supply, voltmeter with digital output(precision, 0.001V), DC amperometer with digital output(precision, 0.001A), calorimeter consisted of electric heater, magnetic stirrer, Beckmann thermometer with digital output, timer, analytical balance, and platform scale.

3.2. Chemicals

KNO_3(AR), $Na_2S_2O_3 \cdot 5H_2O$(AR), KCl(AR).

4. Procedures

Select 4.1 or 4.2 to perform the experiment.

4.1. Determination of integral dissolution heat of KNO_3 by electrothermic method

4.1.1. Keep the pre-grinded KNO_3 in a desiccator.

4.1.2. Weigh out 8 samples of KNO_3 precisely with the analytical balance to about 2.5, 1.5, 2.5, 3, 3.5, 4, 4 and 4.5g respectively and keep them in desiccator before use.

4.1.3. Correctly weigh out 216.2g (12mol) of water into Dewar flask with a platform scale. Record the temperature of the water and take it as the experimental temperature.

4.1.4. Assemble the calorimeter and connect the circuit as illustrated in Fig. 2-2-2. Switch on the magnetic stirrer to a moderate rate. Then switch on the power supply, set the voltage and current to 5V and 0.5A respectively, and record the data precisely. When the digital Beckmann thermometer indicates a 0.5℃ increase of temperature, read it and take it as the starting temperature. Subsequently, press the timer at once to record the time.

4.1.5. Add the first sample of KNO_3, pull out the filler and cover the lid. After the temperature reverts to the starting temperature, record the heating time (remember not to stop the timer). Then add the second sample of KNO_3.

4.1.6. Repeat procedure 4.1.5 till all 8 samples of KNO_3 are added.

4.1.7. Clean the apparatus and sort the experiment table.

4.2. Determination of the integral heat of solution of KNO_3 or $Na_2S_2O_3 \cdot 5H_2O$ by standard substance method

4.2.1. Calibration of capacity of calorimeter

The integral heats of solution of KCl are listed in Appendice Table 4-6. The calorimeter assembly is almost the same as that shown in Fig. 2-2-2 except that the electrothermic system is removed.

Accurately add 360ml of water into a Dewar flask with a 500ml measuring cylinder. After the lid is inserted, the mixture is stirred gently and evenly to reach a temperature balance between the distilled water and the calorimetric system. The temperature of the water is measured by a digital Beckmann thermometer. The starting temperature ($T_{initial}$) is the temperature that keeps almost constant for 5 min.

Add quickly about 7.5±0.01g KCl into Dewar flask. Stir gently and evenly as well after the lid is inserted. Record the temperature every 1min until the temperature is not changed for 5min. Take the temperature as T_{final}. Clean the calorimeter and dry it for later use.

Select either experiment 4.2.2 or experiment 4.2.3.

4.2.2. Integral heat of solution of KNO_3: Weigh out 5.1g of KNO_3(1mol KNO_3 per 400mol water) into 360ml of water and repeat the procedure as described in 4.2.1 to determine the $T_{initial}$ and T_{final} respectively.

4.2.3. Integral heat of solution of $Na_2S_2O_3 \cdot 5H_2O$: Weigh out 12.4g of $Na_2S_2O_3 \cdot 5H_2O$ (1mol $Na_2S_2O_3 \cdot 5H_2O$ per 400mol water) into 360ml of water and repeat the procedure as described in 4.2.1 to determine the $T_{initial}$ and T_{final} respectively.

5. Data recording and processing

5.1. Fill table 2-2-1 with the data measured and calculated.

5.2. Plot $\Delta H_m^{\ominus}$ versus n_0.

5.3. Plot $\Delta H_m^{\ominus}$ versus $\Delta H_m^{\ominus}/n_0$, and calculate ΔH_m^{∞} and b from the intercept and slope of the line respectively.

5.4. Calculate the integral heats of solution of KNO_3 at n_0 = 100mol and n_0 = 200mol. Calculate the integral heat of dilution when n_0 changes from 100mol to 200mol.

5.5. Calculate the differential heat of solution and differential heat of dilution of KNO_3 at n_0 = 200mol.

5.6. Calculated the integral heat of solution of KNO_3 or $Na_2S_2O_3 \cdot 5H_2O$ according to equation(2-2-6).

6. Questions

6.1. Can the electrothermic method be used to measure the heat of solution for an exothermic reaction?

6.2. Can the electrothermic method be used to measure the specific heat effect of a liquid, heat of hydration, heat of formation or heat of mixing?

6.3. Why the temperature of the system should be 0.5℃ higher than that of the surroundings when the experiment is started?

6.4. How to reduce the errors in determination of heat of solution? What factors will influence the result of the experiment?

7. Pharmaceutical applications

Drugs given by solid dosage form must release from the preparation and dissolve under physiological conditions before absorption. Heat generated in the process of dissolution affects greatly on drug dissolution. It was reported that the initial dissolution rate and solubility of the drug were correlated with the heats of solution, which would be applicable for the quality control of the drug. Heat of solution is related to the crystal forms of drugs. It is an indirect measure of the lattice energy of a solid, and is well correlated with the crystallinity of all the solid forms. Therefore, heat of solution is also a method for characterization of pharmaceutical polymorphs.

8. Calibration and measurement of General Beckmann thermometers

In the experiments of physical chemistry, it is usually desired to measure temperature changes precisely. However, a common thermometer is unable to attain at high accuracy, but a Beckmann rage thermometer can be.

A Beckmann thermometer is a device used to measure small differences of temperature at high accuracy, as shown in Fig. 2-2-3. A typical Beckmann thermometer is usually 40 ~ 50cm in length. The temperature scale typically covers about 5℃ and is graduated in units of 0.01℃. With a magnifier it is possible to estimate temperature changes to 0.002℃. The Beckmann thermometer is designed to have the mercury reservoir located at the upper end of the thermometer tube which allows you to adjust the amount of mercury in the thermometer to measure the temperature in different ranges. Two methods to do this are as follows.

(1) Gently warm the bulb of the thermometer in a water bath with high temperature. Then take it out, invert it to allow the mercury in the bulb to join the mercury in the reservoir. It is then turned upright and placed in a bath one or two degrees above the upper limit of temperatures to be measured more than 5min. Take it out, hold it vertically, tap the right arm to separate the mercury at the bend. Place the thermometer in the system to be measured. If the reading rests within the desired scale, the thermometer is ready for use, otherwise repeat the operation until the mercury stands where we want it. (Fig. 2-2-3b)

(2) Use the upper part of Beckman thermometer temperature scale to adjust the amount of mercury. First estimate the highest temperature the thermometer to record in the experiment. Invert the thermometer so that the mercury in the bulb and reservoir will transfer to the end of the capillary. Then slowly tilt the thermometer until the mercury in the reservoir joins the thread in the capillary, as shown in Fig. 2-2-3c. If the estimated temperature is higher than room temperature, make the mercury flow to the reservoir by means of warm water or gravity force until the temperature in the upper scale reaches the expected value, and then slightly tap it so that the mercury thread breaks at the bend. If, on the other hand, the estimated temperature is lower than the room temperature, immerse the thermometer in cold water bath so that the mercury level declines to the required value in the scale, thereby break the mercury in the same way.

Notes: Beckmann thermometer is longer than ordinary glass-thermometer and more prone to damage. It can only be placed in a bath, in the box, or in your hand, but not be placed arbitrarily. Be sure to avoid its use under very high or very low temperature and prevent to shift the mode of base temperature during the experiment.

实验三 凝固点降低法测定摩尔质量

(一) 实验目的

1. 采用凝固点降低法测定萘的摩尔质量。
2. 掌握溶液凝固点的测量技术，并加深对稀溶液依数性的理解。
3. 了解凝固点降低的药学应用。

(二) 实验原理

稀溶液具有依数性，凝固点降低是依数性的一种表现。固体溶剂与溶液保持平衡时的温度称为溶液的凝固点。在溶液浓度很稀时，确定溶剂的种类和数量后，溶剂凝固点降低值仅仅取决于所含溶质分子的数目。

稀溶液的凝固点降低（对析出物为纯固相溶剂的体系）与溶液成分的关系式为

$$\Delta T_f = \frac{R\,(T_f^*)^2}{\Delta H_m} \cdot \frac{n_2}{n_1+n_2} \qquad 式(2\text{-}3\text{-}1)$$

式中 ΔT_f 为凝固点降低值；T_f^* 为以绝对温度表示的纯溶剂的凝固点；ΔH_m 为摩尔熔化热；n_1 为溶剂的物质的量；n_2 为溶质的物质的量。

当溶液很稀时，$n_2 \ll n_1$，则有

$$\Delta T_f = \frac{R\,(T_f^*)^2}{\Delta H_m} \cdot \frac{n_2}{n_1} = \frac{R\,(T_f^*)^2}{\Delta H_m} \cdot M_1 m_2 = K_f m_2 \qquad 式(2\text{-}3\text{-}2)$$

式中 M_1 为溶剂的摩尔质量；m_2 为溶质的质量摩尔浓度；K_f 为溶剂的凝固点降低常数。

如果已知溶剂的凝固点降低常数 K_f，并测得该溶液的凝固点降低值 ΔT_f，控制溶剂和溶质的量为 W_1 和 W_2，就可以通过下式计算溶质的摩尔质量 M_2。即

$$M_2 = K_f \cdot \frac{W_2}{\Delta T_f \cdot W_1} \times 10^3 \qquad \text{式(2-3-3)}$$

凝固点测定方法是将已知浓度的溶液逐渐冷却成过冷溶液，然后促使溶液结晶；当晶体生成时，放出的凝固热使系统温度回升，当放热与散热达成平衡时，温度不再改变，则固液两相达成平衡，此时的温度即为溶液的凝固点。

纯溶剂在凝固前温度随时间均匀下降，当达到凝固点时，固体析出，放出热量，补偿了对环境的热散失，因而温度保持恒定，直到全部凝固后，温度再均匀下降，其冷却曲线见图 2-3-1(a)所示。实际上纯液体凝固时，由于开始结晶出的微小晶粒的饱和蒸气压大于同温度下的液体饱和蒸气压，所以往往产生过冷现象，即液体的温度要降到凝固点以下才析出固体，随后温度再上升到凝固点，见图 2-3-1(b)所示的冷却曲线。

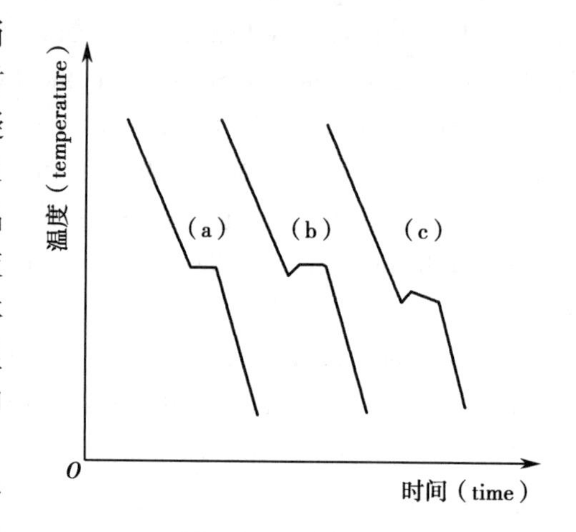

图 2-3-1 冷却曲线
Fig. 2-3-1 Cooling Curves

溶液的冷却情况与此不同，其凝固点低于纯溶剂的凝固点。当溶液冷却到凝固点，开始析出固态纯溶剂。随着溶剂的析出，溶液浓度相应增大，所以溶液的凝固点随着溶剂的析出而不断下降，在冷却曲线上得不到温度不变的水平线段。当有过冷情况发生时，溶液的凝固点应从冷却曲线上待温度回升后外推而得，见冷却曲线图 2-3-1(c)所示。

（三）仪器和试剂

1. 仪器 凝固点测定仪，普通温度计，25ml 移液管，压片机，烧杯，精密温差测量仪。

2. 试剂 环己烷(分析纯)，萘(分析纯)。

（四）实验步骤

1. 仪器安装 按图 2-3-2 所示安装凝固点测定仪，注意测定管，搅拌棒都须清洁、干燥，温差测量仪的探头、温度计都须与搅拌棒有一定空隙，防止搅拌时发生摩擦。

2. 水浴温度调节 调节水浴温度低于环己烷凝固点 2~3℃，并应经常搅拌，不断加入碎冰，使冰浴温度保持基本不变。

3. 温差测量仪示零调节 将测量仪的探头放入测量管中，使数字显示为“0”左右。

4. 环己烷的参考温度测定 准确移取 25ml 环己烷，小心注入测定管中，塞紧软木塞，防止环己烷挥发，记下环己烷的温度值。将测定管直接放入冰浴中，不断移动搅拌棒，使环己烷逐步冷却。当刚有固体析出时，迅速取出测定管，擦干管外冰水，插入空气套管中，缓慢均匀搅拌，每 30 秒读取精密温差测量仪显示的数值，直至温度稳定，即为环己烷的凝固点参考温度。

5. 环己烷的凝固点测定 将测定管取出，用手温热，同时搅拌，使管中固体完全熔化，再将测定管直接插入冰浴中，缓慢搅拌，使环己烷迅速冷却，当温度降至高于凝固点参考温度

0.5℃时，迅速取出测定管，擦干，放入空气套管中，每秒搅拌一次，使环己烷温度均匀下降，当温度低于凝固点参考温度时，应急速搅拌（防止过冷超过 0.5℃），促使固体析出，温度开始上升，搅拌减慢，每隔 30 秒读取温差测量仪的温度示值，直至温度恒定，此即为环己烷的凝固点。重复测定三次，要求环己烷凝固点的绝对平均误差小于±0.003℃。

6. 溶液的凝固点测定 取出测定管，使管中的环己烷熔化，从测定管的支管加入事先压成片状的萘（约 0.2~0.3g），待溶解后，用上述步骤 4 和 5 测定溶液的凝固点。先测凝固点的参考温度，再精确测之。溶液凝固点是取过冷后温度回升所达的最后温度，重复三次，要求绝对平均误差小于±0.003℃。

7. 实验完毕，关闭电源，清洗仪器，整理实验台。

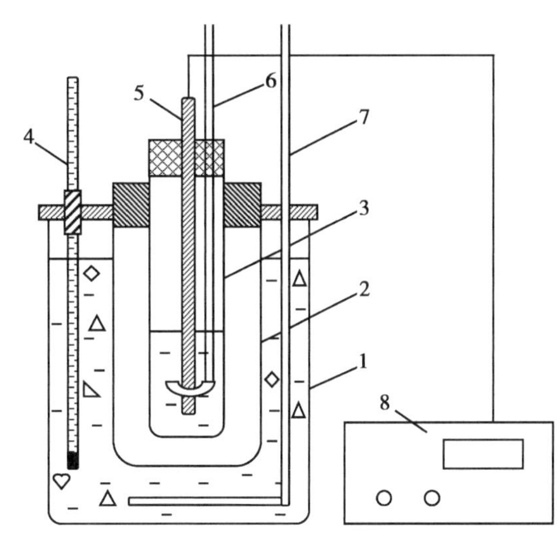

1. 大烧杯（large beaker）；2. 空气套管（air jacket）；3. 带加料口的测量管（test tube）；4. 普通温度计（thermometer）；5. 感温探头（temperature probe）；6. 搅拌器（stirrer）；7. 搅拌器（stirrer）；8. 温差仪（temperature measuring instrument）

图 2-3-2 实验装置

Fig. 2-3-2 Schematic of Experimental Apparatus

8. 注意事项

（1）注意搅拌速度以及加冰的量，以控制过冷程度。

（2）测定管内管、搅拌棒及温度传感器探头需要干燥洁净，防止带入水分。

（五）数据记录与处理

1. 用$\rho(\mathrm{kg/m^3}) = 0.797\ 1\times10^3 - 0.887\ 9t$ 计算室温时环己烷密度，并由此计算所取环己烷的质量 W_1(g)。

2. 由测定的纯溶剂和溶液凝固点 T_f^* 和 T_f 计算萘的摩尔质量。已知环己烷的凝固点 $T_f^* = 279.7\mathrm{K}$，$K_f = 20.1\mathrm{K\cdot kg/mol}$。

（六）思考题

1. 凝固点降低公式的使用条件是什么？为什么必须控制溶质的加入量？

2. 为什么测量溶液凝固点时，必须尽量减少过冷现象？而测量溶剂凝固点时，却没有做此要求？

3. 环己烷挥发和萘含有杂质分别对实验结果有何影响？

（七）药学应用

凝固点降低值的多少，直接反映了溶液中溶质的质点数目。溶质在溶液中的离解、缔合、溶剂化和配合物生成等情况，都会影响凝固点降低值。因此，溶液的凝固点降低法可用于研究药物分子在溶液中的状态。

（八）其他测量方法或系统

1. 由于环己烷和萘具有一定毒性，结合学生实验绿色化，可以改为水和蔗糖（或葡萄糖）系统。

2. 为了进一步控制搅拌速度和系统温度，可以采用自冷式凝固点测定仪。

Experiment 3 Determination of Molar Mass by Freezing Point Depression

1. Objective

1.1. To determine molar mass of naphthalene by freezing point depression method.

1.2. To master experimental technology of freezing point determination and understand deeply the colligative properties of solutions.

1.3. To learn the application of freezing point depression in pharmaceutical science.

2. Principle

The colligative properties of dilute solution are in proportion to the concentration of solute dissolved in solvent. These properties depend only on the number of solute particles. At constant pressure, the pure liquid and pure solid can be in equilibrium at the constant temperature, namely, the freezing point of the compound. When the concentration of the solution is very diluted, after determining the type and quantity of solvent, the solvent freezing point decreases only depending on the number of solute molecules contained.

The change in the freezing point (ΔT_f) for a nonvolatile solvent can be determined using the following equation

$$\Delta T_f = \frac{R\,(T_f^*)^2}{\Delta H_m} \cdot \frac{n_2}{n_1+n_2} \tag{2-3-1}$$

where ΔT_f is the freezing point depression value, T_f^* is the freezing point of pure solvent, ΔH_m is the molar heat of fusion, n_1 and n_2 are amount of substance of solvent and solute, respectively.

When the solution is very dilute ($n_2 \ll n_1$), equation (2-3-1) becomes

$$\Delta T_f = \frac{R\,(T_f^*)^2}{\Delta H_m} \cdot \frac{n_2}{n_1} = \frac{R\,(T_f^*)^2}{\Delta H_m} \cdot M_1 m_2 = K_f m_2 \tag{2-3-2}$$

where M_1 is the molar mass of the solvent, m_2 is the molality of the solute, and K_f is the freezing point depression constant of the solvent.

When the ΔT_f, K_f, weights of solvent W_1 and solute W_2 are obtained, the molar mass of solute M_2 will be calculated using the following equation

$$M_2 = K_f \cdot \frac{W_2}{\Delta T_f \cdot W_1} \times 10^3 \tag{2-3-3}$$

If a pure substance is heated to a liquid state and allowed to cool, the temperature of the liquid will begin to drop. When the freezing point of the substance is reached, the liquid begins to solidify and the temperature of the solid-liquid mixture will remain constant until all the liquid has been converted into solid. Then the temperature will begin to drop again. A graph of the temperature of such a system as function of time can be drawn and is called a cooling curve which exhibits a typical plateau at the freezing point of the substance [Fig. 2-3-1(a)].

The cooling behavior of a solution is somewhat different from that of a pure liquid. The temperature at which the solution begins to solidify is lower (i.e. depressed) than that of the pure solvent. Additionally, there is a slow gradual fall in temperature as freezing proceeds [Fig. 2-3-1 (c)]. The difference between the freezing points of the pure solvent and solution can be used to calculate the molality of the solution or the molar mass of the solute using equation (2-3-2) and equation (2-3-3).

In Fig. 2-3-1(b) and (c), the supercooling effect may be seen. Sometimes, a liquid or solution is solidified below the actual freezing point initially and then come back up to the freezing point temperature as the solid forms. This behavior is commonly seen when very clean or new equipment is used because of lacking scratches or other irregularities that serve as sites for crystallization to begin. Supercooling may be overcome if adequate churning of the sample is provided. When determining the freezing point, the supercooling effect should be ignored if present.

3. Apparatus and chemicals

3.1. Apparatus

Freezing-point measurement instrument, thermometer, 25ml pipette, tablet machine, precision temperature measuring instrument.

3.2. Chemicals

Cyclohexane (AR), naphthalene (AR).

4. Procedures

4.1. Assemble the apparatus as shown in Fig. 2-3-2 and adjust so that the stirrer, thermometer and probe will not touch each other and not touch the bottom and wall of the test tube. The test tube, probe and stirrer inside must be clean and dry.

4.2. Regulate the temperature 2~3℃ lower than the freezing point of cyclohexane and keep it basically unchanged through stirring and ice addition.

4.3. Place the temperature probe into the test tube and regulate the temperature measuring instrument to display a value as zero.

4.4. Determination of the reference freezing point of cyclohexane

Measure precisely 25ml of cyclohexane with the pipette and pour it into a dry test tube. Stop the test tube with a rubber stopper to prevent evaporation of cyclohexane. Record the initial temperature of the cyclohexane. Place the test tube directly into ice-water bath with only the bottom of the tube in the bath. Keep stirring gently until the solid begins to appear. Remove the test tube from the ice-water bath and dry the outside. Insert the test tube into the air jacket and stir the cyclohexane slowly and continuously. Record the temperature each 30s until a constant temperature is obtained. Take this temperature as the reference freezing point of cyclohexane.

4.5. Determination of the freezing point of cyclohexane

Remove the test tube from the air jacket and warm it up till all solid has melted. Replace the test tube into the ice-water bath to cool the cyclohexane. Once the thermometer displays a temperature of approximately 0.5℃ higher than the reference freezing point of cyclohexane, remove the test tube from the bath and dry the outside. Place the test tube into the air jacket again and stir it slowly. When the temperature is lower than the reference freezing point, stir quickly to prevent a supercooling over 0.5℃. When the temperature begins to increase, stir slowly and record the temperature each 30s until a constant temperature is obtained. This is the freezing point of cyclohexane. Repeat the cooling process three times and the absolute average error must be less than 0.003℃. Save the test tube of cyclohexane for the next procedure.

4.6. Determination of the freezing point of the solution

Remove the test tube and allow it to warm until all the cyclohexane is melted. Add a sample tablet(about 0.2 ~ 0.3g) into the test tube and stir the solution until the solid is completely dissolved. Determine the freezing point of the solution using the same procedures described in 4.4 and 4.5. Repeat the cooling process three times and the absolute average error must be less than 0.003℃.

4.7. Dispose of the solution into the labeled waste container in the hood. Rinse the test tube, probe and stirrer with acetone and dispose these rinses in the waste bottle.

4.8. Attentions

4.8.1. The stirring speed and the amount of ice should be controlled to avoid too high degree of supercooling.

4.8.2. Experimental tube, stirring rod and temperature sensor probe must be dry and clean to prevent the water into test system.

5. Data recording and processing

5.1. Calculate the density of cyclohexane at room temperature using the equation $\rho(kg/m^3) = 0.797\ 1\times10^3-0.887\ 9t$, and then calculate the mass $W_1(g)$ of cyclohexane to be taken.

5.2. Calculate the molar mass of naphthalene using equation (2-3-3). ($T_f^* = 279.7K$, $K_f = 20.1K \cdot kg/mol$)

6. Questions

6.1. What is the condition for use of the equation to calculate the freezing point depression? Why the amount of solute should be controlled?

6.2. Why the degree of supercooling should be controlled when the freezing point of the solution is measured, but no such requirement is made for determination of that of solvent?

6.3. What would be the effect on the calculated molar mass if some impurity were present in naphthalene or some of the cyclohexane evaporated before the solute was added?

7. Pharmaceutical applications

The amount of freezing point depression value directly reflects the number of particles of solute in solution. When there is dissociation, association, solvation or formation of complex, etc., the freezing point will be affected. Therefore, the freezing point depression of the solution can be used to study the state of drug molecules in solution.

8. Other methods or systems

8.1. Both the cyclohexane and naphthalene are toxicity. Replace the cyclohexane and naphthalene by water and sucrose (or glucose) in consideration of green chemical experiment.

8.2. For better control of the stirring speed and the system temperature, the self-cooling freezing point measuring instrument can be used.

实验四 凝固点降低法测定氯化钠注射液的渗透压

（一）实验目的及要求

1. 掌握凝固点降低法测定溶液渗透压的原理。
2. 掌握溶液凝固点的测定技术，并加深对稀溶液依数性的理解。
3. 了解渗透压测定的药学应用。

（二）实验原理

溶剂通过半透膜由低浓度溶液向高浓度溶液扩散的现象称为渗透，阻止渗透所需施加的压力，即渗透压。渗透压是稀溶液的依数性之一，其值依赖于溶液中溶质粒子的数量，通常以渗透压摩尔浓度表示，它反映的是溶液中各种溶质对溶液渗透压贡献的总和。每千克溶剂中溶质的毫渗透压摩尔，称为毫渗透压摩尔浓度（mOsmol/kg），其计算公式为

$$\text{毫渗透压摩尔浓度} = mi \times 1\,000 \qquad \text{式(2-4-1)}$$

式中 m 为质量摩尔浓度，即每千克溶剂中溶解溶质的摩尔数；i 为一个溶质分子溶解并解离时形成的粒子数。在理想溶液中，例如葡萄糖 $i=1$，氯化钠或硫酸镁 $i=2$，氯化钙 $i=3$，枸橼酸钠 $i=4$。在真实溶液中，由于分子缔合等作用，溶液的理论渗透压摩尔浓度不容易计算，因此通常采用实际测定值表示。

凝固点降低也是稀溶液依数性的一种表现。固体溶剂与溶液平衡时的温度称为溶液的凝固点。在溶液浓度很稀时，确定了溶剂的种类和数量后，溶剂凝固点降低值仅取决于所含溶质分子的数目。

实验三的实验原理中已经给出，对很稀的溶液（$n_2 \ll n_1$），凝固点降低与溶质的质量摩尔浓度之间的关系为

$$\Delta T_{\mathrm{f}} = \frac{R\,(T_{\mathrm{f}}^{*})^2}{\Delta H_{\mathrm{m}}} \cdot \frac{n_2}{n_1} = \frac{R\,(T_{\mathrm{f}}^{*})^2}{\Delta H_{\mathrm{m}}} \cdot M_1 m_2 = K_{\mathrm{f}} m_2 \qquad \text{式(2-4-2)}$$

式中 ΔT_{f} 为溶液的凝固点降低值；T_{f}^{*} 为纯溶剂的凝固点；ΔH_{m} 为摩尔熔化热；n_1 为溶剂的物质的量；n_2 为溶质的物质的量；M_1 为溶剂的摩尔质量；m_2 为溶质的质量摩尔浓度；K_{f} 为溶剂的凝固点降低常数。

溶液的渗透压常以质量渗透摩尔浓度来表示。因此，可以用凝固点下降原理测定溶液的渗透压，计算公式为

$$O_{\mathrm{s}} = 1\,000 \times \frac{\Delta T}{K_{\mathrm{f}}} \qquad \text{式(2-4-3)}$$

如实验三中所述，将已知浓度的溶液逐渐冷却至某一温度，使溶剂凝固析出，通过作冷却曲线可以确定凝固点。实际测定时往往会产生过冷现象，溶剂和溶液凝固点的确定如图 2-4-1（a）和（b）所示。

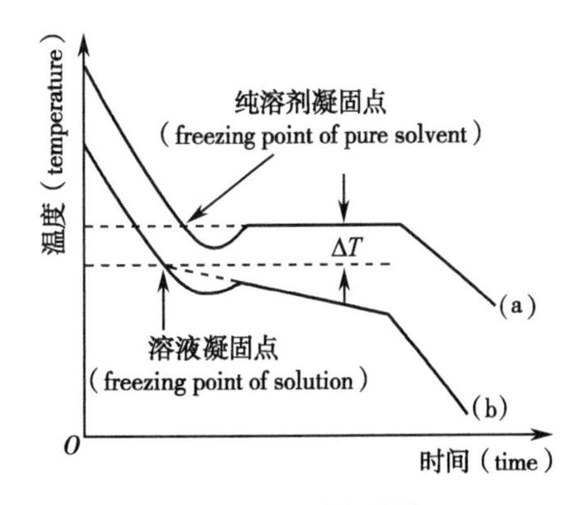

图 2-4-1 冷却曲线

Fig. 2-4-1 Cooling Curves

（三）仪器和试剂

1. 仪器 凝固点测定仪，贝克曼温度计，普通温度计，25ml 移液管，烧杯。

2. 试剂 水，1.0% 氯化钠标准溶液，0.9% 氯

化钠注射液,5% 葡萄糖注射液,粗盐。

(四) 实验步骤

1. 仪器安装 图 2-4-2 为仪器装置示意图。将盛有待测溶液的测定管(内管),隔着空气套管放在冰浴中,在测定管中放贝克曼温度计传感器和玻璃搅拌棒,冰浴中插入温度计和搅拌棒,用于指示冰浴温度及冰浴搅拌。仪器安装时,注意测定管中的液面须在冰浴液面之下,测定管和搅拌棒都须清洁、干燥,温差测量仪的探头、温度计都须与搅拌棒有一定空隙,防止搅拌时发生摩擦。

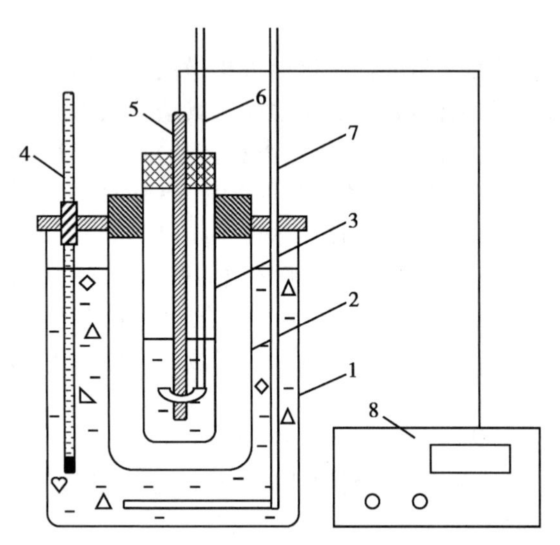

1. 大烧杯(large beaker);2. 空气套管(air jacket);3. 带加料口的测量管(test tube);4. 普通温度计(thermometer);5. 感温探头(temperature probe);6. 搅拌器(stirrer);7. 搅拌器(stirrer);8. 温差仪(temperature measuring instrument)

图 2-4-2 实验装置

Fig. 2-4-2 Schematic of Experimental Apparatus

2. 冰浴温度调节 冰浴中放入适量的冷水、碎冰和食盐,使冰浴温度低于冰点 2~3℃,并应经常搅拌,不断加入碎冰,使冰浴温度保持基本不变。

3. 贝克曼温度计的校正 量取约 25ml 的 1.0% 氯化钠标准溶液于测定管中,插入贝克曼温度计传感器,置于冰水浴中,不断上下搅拌,同时记录温度随时间的变化,绘制冷却曲线,并按图 2-4-1 的方法读取凝固点。同法测定纯水的凝固点,由这两个数据对温度计进行校正。

4. 样品的测定

(1) 取 0.9% 氯化钠注射液约 25ml 于测定管中,插入贝克曼温度计传感器,置于冰水浴中,不断上下搅拌,同时记录温度随时间的变化,绘制冷却曲线,并按图 2-4-1 的方法读取凝固点,计算凝固点下降值。

(2) 测定 5% 葡萄糖注射液的凝固点,方法同 0.9% 氯化钠。

5. 实验完毕,关闭电源,清洗仪器,整理实验台。

(五) 数据记录与处理

1. 已知 1.0% 氯化钠标准溶液的凝固点 $T_f^* = 0.58$℃,水的凝固点 $T_f^* = 0$℃,$K_f = 1.86\text{K}\cdot\text{kg/mol}$。

2. 按表 2-4-1 和表 2-4-2 及公式(2-4-3)记录并处理数据。

表 2-4-1 温度计校正数据记录表

Table 2-4-1 Data Sheet for Thermometer Calibration

溶液(solution)	c	T_f/℃	ΔT_f/℃	温度计校正(thermometer calibration)
标准溶液 1(standard 1)				
标准溶液 2(standard 2)				

表 2-4-2 凝固点降低及渗透压数据处理表

Table 2-4-2 Data Sheet for Calculation of Freezing Point Depression and Osmosis Pressure

溶液(solution)	T_f/℃	ΔT_f/℃	mOsmol/kg
0.9% 氯化钠注射液 (0.9% NaCl injection)			
5% 葡萄糖注射液 (5% glucose injection)			

(六)思考题

1. 过冷现象是如何产生的?如何控制过冷程度?
2. 溶质的加入量遵循什么原则?加入量太多或太少对结果影响如何?

(七)药学应用

生物膜,例如人体的细胞膜或毛细血管壁,一般具有半透膜的性质,在涉及溶质扩散或通过生物膜的液体转运的各种生物过程中,渗透压都起着极其重要的作用。因此渗透压或渗透压测定对人体内器官和组织的生理功能和新陈代谢活动,对注射剂、滴眼剂、输液等剂型的制备等具有重要意义。《中华人民共和国药典》2020 年版收录了渗透压摩尔浓度测定方法,对相关药品的渗透压进行检测,为规范药品生产提供了检测依据。目前,除了最常用的冰点下降法外,沸点升高法、蒸气压下降法和半透膜法也可以用来测定样品的渗透压。

Experiment 4 Determination of Osmotic Pressure of Sodium Chloride Injection by Freezing Point Depression

1. Objective

1.1. To learn the principle to determine the osmotic pressure of the solution by freezing point depression method.

1.2. To master experimental technology of freezing point determination and understand deeply the colligative properties of diluted solutions.

1.3. To learn the pharmaceutical application of determination of osmotic pressure.

2. Principle

Osmosis is the process whereby a pure liquid passes through a semi-permeable membrane from low concentration to high concentration. Osmotic pressure is the pressure that must be applied on the high concentration side to stop osmosis. Osmotic pressure is a colligative property of dilute solutions which depends on the concentration of dissolved particles. Osmotic pressure is usually expressed in terms of osmolality, which refers to the number of osmole of solute per kilogram of solvent (Osmol/kg). Osmolality can also be expressed as milliosmole per kilogram solvent (mOsmol/kg) and be calculated as

$$\text{mosmole} = mi \times 1\ 000 \qquad (2\text{-}4\text{-}1)$$

where m is molality, defined as the moles of solute per kilogram of solvent, and i is the number of particles generated in per mole of solute (e.g., $i = 1$ for glucose; $i = 2$ for NaCl or $MgSO_4$; $i = 3$ for $CaCl_2$; $i = 4$ for sodium citrate when they are expected as ideal solution or ideal strong electrolytes). Due to interionic forces, molecule association etc. in real solution, osmolality fails to be calculated theoretically but experimentally measured.

Freezing point depression is also a colligative property. The freezing point of a solution is the temperature at which solid solute and solution coexist in equilibrium. When the concentration of the solution is very diluted, after determining the type and quantity of solvent, the solvent freezing point reduction value depends only on the number of solute molecules contained.

As discussed in experiment 3, the freezing point depression (ΔT_f) for a very dilute solution, moles of solute (n_2) can be neglected when compared with moles of solvent (n_1), can be determined using the following equation

$$\Delta T_f = \frac{R\,(T_f^*)^2}{\Delta H_m} \cdot \frac{n_2}{n_1} = \frac{R\,(T_f^*)^2}{\Delta H_m} \cdot M_1 m_2 = K_f m_2 \qquad (2\text{-}4\text{-}2)$$

where ΔT_f is the freezing point depression of the solution, T_f^* is the freezing point of pure solvent, ΔH_m is the molar heat of fusion of solvent, M_1 is the molar mass of the solvent, m_2 is the total molality of the solute, and K_f is the freezing point depression constant of the solvent.

As mentioned previously, osmotic pressure is also commonly demonstrated as milliosmole per kilogram solvent, so it is possible to calculate the osmotic pressure from freezing point depression ΔT_f according to the following equation

$$O_s = 1\ 000 \times \frac{\Delta T}{K_f} \qquad (2\text{-}4\text{-}3)$$

As expressed in experiment 3, the freezing point can be determined from the cooling curves when a solution of known concentration gradually cools to a certain temperature. However during actual determination, a phenomenon called supercooling usually occurs. Therefore, the determination of the freezing points for both the pure solvent and a tested solution are shown in Fig. 2-4-1(a) and Fig. 2-4-1(b).

3. Apparatus and chemicals

3.1. Apparatus

Freezing-point apparatus, Beckmann thermometer, common thermometer, 25ml pipette, beaker.

3.2. Chemicals

Water, 1.0% of sodium chloride standard solution, 0.9% of sodium chloride injection, 5% of glucose injection, salt.

4. Procedures

4.1. Assemble the apparatus as shown in Fig. 2-4-2. Be sure to use a clean and dry test tube, thermometer probe and stirrer inside. Place the test tube, thermometer probe and stirrer into ice bath. The level of the water in the test tube should be below the level of the ice bath. Note that do not make the thermometer and stirrer inside touching each other as well as touching the bottom and wall of the test tube.

4.2. Pour cold water, ice and salt into the ice bath to create an ice-salt water bath at the temperature 2~3℃ below the freezing point of water. Keep it basically unchanged through stirring and ice addition.

4.3. Calibration of thermometer

Take about 25ml of water and 1.0% of sodium chloride standard solution respectively into the test tubes. Place the thermometer probe into the test tube. Fix the test tube in the ice-salt water bath and keep stirring. Record the temperature until the sample totally solidified. Draw the cooling curves to obtain the freezing points of water and 1.0% of sodium chloride standard solution according to Fig. 2-4-1. Calibrate the Beckmann thermometer using the freezing points measured.

4.4. Sample determination

4.4.1. Take about 25ml of 0.9% of sodium chloride injection into the test tube. Place the thermometer probe into the test tube. Fix the test tube in the ice-salt water, record the temperature of the sample with time at stirring until all the solution solidifies. Draw the cooling curve to obtain the freezing point of the sample.

4.4.2. The freezing point depression of 5% of glucose injection is determined by the same method as 4.4.1.

4.5. Once the runs has finished, turn off the power supply, clean the lab glassware and other equipment. Tidy the experimental desk.

5. Data recording and processing

5.1. From the cooling curves of water, determine the freezing point of each sample and calculate the freezing point depression of 1.0% of sodium chloride standard solution, 0.9% of sodium chloride injection and 5% of glucose injection. For the 1.0% of sodium chloride standard solution, $T_f^* = 0.58℃$. For solvent water, $K_f = 1.86K \cdot kg/mol$.

5.2. Fill table 2-4-1 and table 2-4-2 with the data measured and calculated.

6. Questions

6.1. What is the explanation of the formation of supercooling phenomena? How to control the degree of supercooling?

6.2. What would be the effect on the experimental result if too much or too less solute was added?

7. Pharmaceutical applications

Important examples of semi-permeable membranes are biological membrane, for example the cell membrane and capillary wall of human body. It is important to regulate osmotic pressure of injection, eye drops, and transfusion etc. So osmotic pressure regulation or determination is important for normal physiological functions and metabolism, or in the manufacture of formulations of injections, eye drops, and transfusion etc. The *Chinese pharmacopoeia* (2020 edition) collects osmolality measurements based on freezing point depression method to regulate drug production. Besides the monitoring freezing point depression, vapor pressure lowering, boiling point elevation and semi-permeable membrane are also used for the determination of osmotic pressure.

实验五 静态法测定液体饱和蒸气压

(一) 实验目的

1. 了解用静态法(等位法)测定水在不同温度下蒸气压的原理,理解纯液体饱和蒸气压与温度的关系。

2. 掌握真空泵、恒温槽及气压计的使用方法。

3. 学会用图解法求所测温度范围内的平均摩尔气化热。

4. 了解饱和蒸气压的药学应用。

(二) 实验原理

一定温度下,密闭于真空容器中的液体与它的蒸气建立动态平衡时,蒸气分子向液面凝结的速度与液体分子从表面蒸发的速度相等,此时液面上的蒸气压力就是液体在该温度下的饱和蒸气压。液体的蒸气压与温度有关,温度升高,分子运动加剧,单位时间内从液面逸出的分子数增多,蒸气压增大;反之,温度降低时,则蒸气压减小。当蒸气压与外界压力相等时,液体沸腾。外压不同时,液体的沸点也不同。当外压为 101.325kPa 时,液体沸腾的温度定为液体的正常沸点。液体的饱和蒸气压与温度的关系可用克劳修斯-克拉珀龙方程式表示,即

$$\frac{\mathrm{d}\ln p}{\mathrm{d}T}=\frac{\Delta_{\mathrm{vap}}H_{\mathrm{m}}}{RT^2} \qquad 式(2\text{-}5\text{-}1)$$

式中 p 为液体在温度 T 时的饱和蒸气压;T 为热力学温度;$\Delta_{\mathrm{vap}}H_{\mathrm{m}}$ 为液体的摩尔气化热;R 为气体常数。当温度变化范围较小时,$\Delta_{\mathrm{vap}}H_{\mathrm{m}}$ 可视为常数,将式(2-5-1)积分,可得

$$\ln p=-\frac{\Delta_{\mathrm{vap}}H_{\mathrm{m}}}{RT}+C \qquad 式(2\text{-}5\text{-}2)$$

式中 C 为积分常数,与压力 p 的单位有关。在一定温度范围内,测定不同温度下的饱和蒸气压,以 $\ln p$ 对 $\frac{1}{T}$ 作图可得一条直线,由直线斜率可以求出摩尔气化热。

静态法测蒸气压的方法是调节外压以平衡液体的蒸气压,求出外压就能直接得到该温度下的饱和蒸气压。

(三) 仪器和试剂

1. 仪器 数字式低真空测定仪,真空泵及附件,恒温槽,温度计,DP-A 精密数字压力计,大气压力计。

2. 试剂 蒸馏水。

(四) 实验步骤

1. 仪器安装 如图 2-5-1 安装各种仪器。其中,由三个相连的玻璃管 a、b 和 c 组成的是平衡管,如图 2-5-2 所示。a、b 和 c 管中装入的均为待测液体,连通后构成 U 型压力计。一定温度下,当 a、c 管的上部充满待测液体的蒸气,b 和 c 管中的液体在同一水平面时,则 c 管液面上的蒸气压与 b 管液面上的压力相等,其值即为待测液体在测定温度下的蒸气压。平衡管与冷凝管用玻璃磨口(或胶塞)密闭相连,以防系统漏气。平衡管中的液体可按如下方

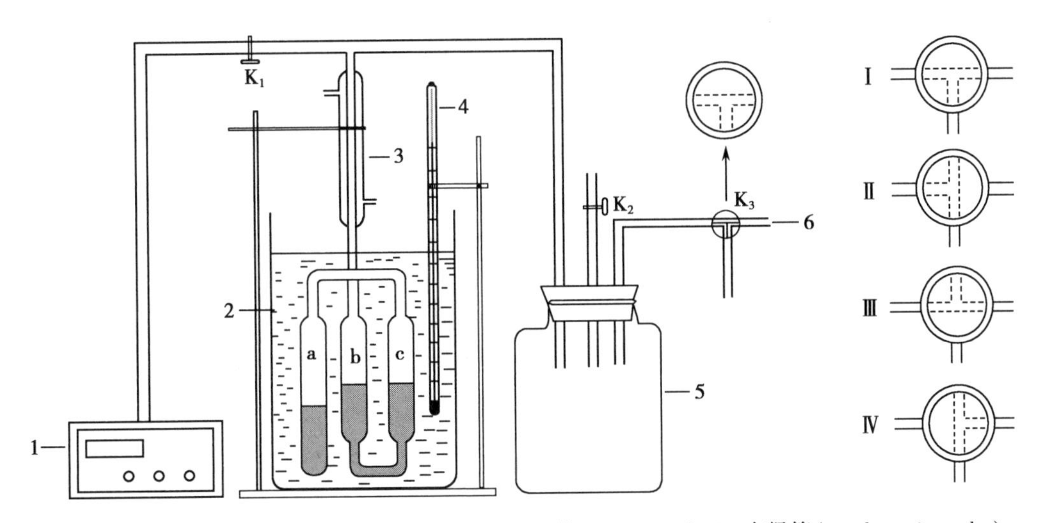

1. 数字式真空测定仪(digital vacuum tester);2. 恒温槽(thermostat);3. 冷凝管(condensation tube);
4. 温度计(thermometer);5. 缓冲瓶(amortization flask);6. 接真空泵(to connect with vacuum pump)

图 2-5-1 蒸气压测定装置图

Fig. 2-5-1 Sketch of Apparatus for Vapor Pressure Determination

法装入:从 b 管的管口加入液体,将平衡管按图 2-5-1 连接,抽气,将系统的压力降低至 50~60kPa,缓慢打开缓冲瓶连通大气阀门 K_3 至Ⅰ位置,大气压力可将液体压入 a 管。

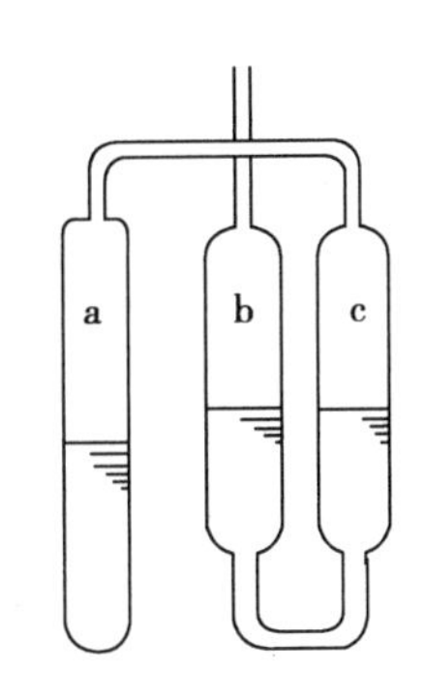

图 2-5-2 平衡管

Fig. 2-5-2 Balancing Tube

2. 检查系统密闭性 打开真空泵,关闭缓冲瓶连通大气的阀门 K_2,K_3 旋至Ⅲ位置,使系统压力降低至 50kPa 左右,将真空泵与缓冲瓶间的进气阀 K_3 右旋至Ⅳ位置,关闭真空泵(注意:只有在Ⅰ或Ⅳ位置时才能关泵)。观察 DP-A 精密数字压力计读数是否发生改变,以检查系统是否漏气。若压力计读数发生明显变化,说明系统漏气,按分段检查法检查堵漏,直至整个系统密闭为止。

3. 水的蒸气压测定 将平衡管全部浸入恒温槽水浴中,同时打开恒温槽开关,接通冷凝水,启动真空泵,将 K_3 右旋至Ⅲ位置,抽净 a 与 c 管上方的空气(此时,应注意调节缓冲瓶上方的平衡阀 K_2 以防止发生暴沸)。当水浴温度达到 40℃时,停止加热。将 K_3 右旋至Ⅳ位置,关闭真空泵。在该温度下恒温 10 分钟后,缓慢调节缓冲瓶上方的平衡阀 K_2 使空气进入体系(注意防止空气进入 a、c 管上方空间,否则需要重新抽净)。当空气将 b 管液面压至与 c 管液面相平时,关闭该平衡阀 K_2,迅速记录温度与压力,此时 b 管液面上方压力即为测量温度下水的饱和蒸气压。同法,依次测定 45℃、50℃、55℃及 60℃时水的饱和蒸气压。

4. 关泵方法 测量结束后,将 K_3 右旋至Ⅰ的位置(防止真空泵油倒吸),关泵。

5. 关闭电源,关闭循环水。读取当天的大气压力($p_{大}$)和室温。

6. 清洗仪器,整理实验台。

7. 注意事项

(1) 注意关泵前三通的位置,将真空泵与缓冲瓶间的进气阀 K_3 右旋至Ⅳ位置,关闭真空泵。

（2）将平衡管全部浸入恒温槽水浴中，保证恒温。

（3）调节平衡时，注意空气不能倒灌。

（五）数据记录与处理

1. 实验数据记录（表 2-5-1）

表 2-5-1 不同温度下系统的压力和液体的饱和蒸气压

Table 2-5-1 System Pressure and Saturated Vapor Pressure of Liquid at Different T

大气压（atmospheric pressure）：________ kPa

温度（temperature）			系统压力（system pressure）/（kPa）$p_{系}=p_{大气}-p_{真空}$	饱和蒸气压（saturated pressure）/（kPa）
T/℃	T/K	$1/T$/（1/K）		
…				
…				
…				

2. 以 $\ln p$ 对 $1/T$ 作图，由该直线斜率计算水的平均摩尔气化热。

（六）思考题

1. 能否在加热条件下检查系统的密闭性？为什么？

2. 为什么要抽除平衡管 a、c 间的空气？

3. 实验过程中为什么要防止 b 上方的空气倒流至 c、a 的上方？如何防止？

4. 缓冲瓶的作用是什么？如果不加缓冲装置，会出现什么现象？

5. 水的平均摩尔气化热为 40.60kJ/mol，根据实验数据计算相对误差，并分析产生误差的原因。

（七）药学应用

饱和蒸气压是液体或固体的非常重要的物理化学性质，在药学中应用广泛。在药物分析中，饱和蒸气压是气相色谱分析顶空技术测定药物中溶剂残留的基础。在药物制剂中，应用饱和蒸气压原理制备气雾剂，特别是通过了解不同温度下的饱和蒸气压，控制常温下气雾剂罐中的压力，以防止意外爆炸。饱和蒸气压在药物合成的分离和提取中也有重要应用，对于高沸点、热不稳定且水不溶性的组分，可通过水蒸气蒸馏在相对较低温度下将其蒸出，再经油水分离可得纯组分，其效率就是由该组分的饱和蒸气压和水的饱和蒸气压之比决定的。利用不同温度下的饱和蒸气压计算出的相变热，对了解药物的溶解、药物的物理稳定性等方面也具有重要意义。

（八）其他测定方法

1. 动态法 测量沸点随施加的外压而变化的一种方法。液体上方的总压力可调节，且用一个大容量的缓冲瓶维持给定值，汞压力计测量压力值，加热液体至沸腾时测量其温度。

2. 饱和气流法 在一定温度和压力下，用干燥气体缓慢地通过被测纯液体，使气流为该液体的蒸气所饱和。用吸收法测量蒸气量，进而计算出蒸气分压，此即该温度下被测纯液体的饱和蒸气压。

Experiment 5 Saturated Vapor Pressure of Pure Liquids by Static Method

1. Objective

1.1. To comprehend the static method of determination of vapor pressure for water at different temperatures, to understand the relationship between saturated vapor pressure and temperature for a pure liquid.

1.2. To learn the approach and the basic operation of vacuum system, thermostat and air barometer.

1.3. To calculate the average molar heat of vaporization in a certain temperature range using graphical method.

1.4. To learn the pharmaceutical applications of vapor pressure.

2. Principle

The saturated pressure of a pure liquid is defined as the pressure of the vapor existing in equilibrium with the liquid sealed in a vacuum container at a certain temperature. When the temperature is raised, the vapor pressure of a liquid increases, because more molecules gain sufficient kinetic energy to break away from the surface of the liquid. Otherwise, the vapor pressure of the liquid decreases with the temperature. When the vapor pressure becomes equal to the atmospheric pressure, the liquid boils, and the temperature is called the boiling point. When the external pressure is 101.325kPa, the boiling temperature of the liquid is called the normal boiling point.

According to the Clausius-Clapeyron equation, variation of the vapor pressure with temperature is given by

$$\frac{\mathrm{d}\ln p}{\mathrm{d}T}=\frac{\Delta_{\mathrm{vap}}H_{\mathrm{m}}}{RT^2} \qquad (2\text{-}5\text{-}1)$$

where p is the saturated vapor pressure of the pure liquid at temperature T, $\Delta_{\mathrm{vap}}H_{\mathrm{m}}$ is the molar heat of vaporization, and R is the gas constant. Within a moderate range of temperature, $\Delta_{\mathrm{vap}}H_{\mathrm{m}}$ could be considered as a constant. Integrate the equation 2-5-1 to yield

$$\ln p=-\frac{\Delta_{\mathrm{vap}}H_{\mathrm{m}}}{RT}+C \qquad (2\text{-}5\text{-}2)$$

where C is the integral constant depending on the unit of pressure. Equation 2-5-2 shows a linear relationship between $\ln p$ and $1/T$. $\Delta_{\mathrm{vap}}H_{\mathrm{m}}$ of the liquid can be estimated from the slope of the straight line.

In the experiment, degree of vacuum, the difference between atmospheric pressure and systemic pressure, is measured instead of systemic pressure.

3. Apparatus and chemicals

3.1. Apparatus

Digital vacuum meter, vacuum pump, thermostat, thermometer, DP-A precise digital pressure gauge, atmospheric pressure gauge.

3.2. Chemicals

Distilled water.

4. Procedures

4.1. Instrument assembly

A sketch of the experiment apparatus is shown in Fig. 2-5-1. Balancing tube (Fig. 2-5-2) is consisted of tris-consecutive glass tubes a, b and c. The tubes are stored with liquid to measure. Tubes b and c are connected to form a U-model pressure gauge. At a specified temperature, when the upper parts of tubes a and c is full of the vapor of measured liquid, the water levels of tubes b and c are the same, indicating the vapor pressure in space of tube c to be equal to the pressure in space of tube b, which is the vapor pressure of the measured liquid. Add the liquid from the opening of tube b, pump down the system until the pressure of the system reaches 50 ~ 60kPa. Open the stopcock K_3 in position Ⅰ, the liquid will be pressed into a tube under the action of atmospheric pressure.

4.2. Detection of system leakage

Open the vacuum pump, close stopcock K_2, turn stopcock K_3 to position Ⅲ, and then lower the system pressure to about 50kPa. Next, turn K_3 to position Ⅳ, and close the vacuum pump. (Attention: the vacuum pump can only be stopped when K_3 is positioned at Ⅰ or Ⅳ.) Observe whether the reading on DP-A precise digital pressure gauge is changed. If the reading changes significantly, it is necessary to examine the location of air leakage until the system is closed tightly.

4.3. Measurement of vapor pressure of water

The cooling water for the reflux head is turned on. Turn on the thermostatic bath and immerse the balancing tube into it. Open the vacuum pump, turn stopcock K_3 to position Ⅲ, and exhaust the air in tubes a and c (Attention: adjust stopcock K_2 in order to prevent liquid flooding). When the temperature of the water reaches 40℃, stop heating. Turn stopcock K_3 to position Ⅳ and stop the vacuum pump. Keep the temperature constant for 10min, and then adjust stopcock K_2 slowly to have air entering the system (air is not allowed to enter into the space of tube a or tube c, otherwise you have to exhaust the air). When the liquid level of tube b is equal to that of tube c, turn off stopcock K_2, and record the temperature and pressure. The pressure above tube b is the saturated vapor pressure of water at a certain temperature. Adjust the system temperature to 45℃, 50℃, 55℃ and 60℃ in turn, and repeat the procedure to measure the saturated vapor pressure at these specified temperatures.

4.4. Turn K_3 to position Ⅰ and stop the vacuum pump.

4.5. Turn off the power supply and circulation water. Read and record the atmospheric pressure and room temperature.

4.6. Clean apparatus, tidy experiment table.

4.7. Attentions

4.7.1. Pay attention to the location of the three-direct-links before turning off the pump. Rotate the inlet valve K_3, which is between the vacuum pump and surge flask, to Ⅳ site. And then turn off the vacuum pump.

4.7.2. Immerse the balance tube into thermostat water bath completely to ensure that the temperature remains constant.

4.7.3. Notice that the air is not expected to flow back during the process of balance regulation.

5. Data recording and processing

5.1. Fill table 2-5-1 with the data measured and calculated.

5.2. Plot $\ln p$ versus $1/T$ to calculate the average molar heat of vaporization from the slope of the line.

6. Questions

6.1. Can the system leakage be checked on condition of heating? Give the reason to your answer.

6.2. Why does the air in tubes a and c have to be exhausted?

6.3. Why the air above tube b is prevented from refluxing into tubes a or c? How to prevent?

6.4. What is the role of surge flask? What will happen without buffer device?

6.5. The average molar heat of evaporation of water is 40.60 kJ/mol. Calculate the relative error according to the data of experiment and discuss the source of errors.

7. Pharmaceutical Applications

Saturated vapor pressure of liquid or solid is a very important physical and chemical property, which is used widely in pharmacy. In pharmaceutical analysis, it is the basis of determination of residual solvents by headspace technique in gas chromatography. In pharmaceutical preparation, the principle of saturated vapor pressure is used to prepare aerosol, especially used to control the pressure of aerosol container at normal temperature to prevent accidental explosion. The saturated vapor pressure is also important in the isolation and extraction of pharmaceutical synthesis. The water insoluble components with high boiling point and thermal instability can be vaporized at relatively low temperature through steam distillation and then be isolated through oil-water separation. The efficiency of this method depends on the ratio between the saturated vapor pressure of the components and the water. What's more, the heat of phase transition obtained from the saturated vapor pressure at different temperatures is essential for us to know about the solubility and the physical stability of drugs.

8. Other methods

8.1. Dynamic method

It is a method for measuring boiling point under different external pressure. The total pressure above liquid is adjustable and a large capacity buffer bottle is used to maintain the given value. Pressure value is measured by mercury manometer and temperature is measured when liquid is heated to boiling.

8.2. Gas saturation method

At specified temperature and pressure, a stream of dry carrier gas is passed over the pure tested liquid in such a way that it becomes saturated with its vapor and the vapor is collected in a suitable trap. Measurement of the amount of material transported by a known amount of carrier gas is used to calculate saturated vapor pressure of the tested pure liquid at a given temperature.

实验六 反应平衡常数及分配系数的测定

（一）实验目的

1. 测定碘和碘离子反应的平衡常数。
2. 测定碘在 CCl_4 和水中的分配系数。
3. 了解温度对平衡常数及分配系数的影响。
4. 了解分配系数的药学应用。

（二）实验原理

1. 分配系数的测定　碘既能溶于四氯化碳也能溶于水，往碘的四氯化碳饱和溶液中加入水，充分振荡，在温度和压力一定时，碘在四氯化碳和水中的分配达到平衡，即

$$I_2(\text{水层}) \rightleftharpoons I_2(\text{四氯化碳层})$$

用硫代硫酸钠标准溶液可以滴定出水层和四氯化碳层中碘的浓度，从而可求出分配系数 K：

$$K=\frac{c_{I_2}(CCl_4\text{ 层})}{c_{I_2}(H_2O\text{ 层})} \qquad \text{式(2-6-1)}$$

2. 化学平衡常数的测定　在一定温度和压力下，碘和碘化钾在水溶液中可建立如下平衡，即

$$I_2+KI \rightleftharpoons KI_3$$

$$K_c=\frac{[KI_3]}{[I_2][KI]} \qquad \text{式(2-6-2)}$$

为测定平衡常数，需测定出平衡时 KI_3、I_2、KI 的浓度，但在上述平衡达到时，若用 $Na_2S_2O_3$ 标准溶液滴定溶液中 I_2 的浓度，则随着 I_2 的消耗，平衡不断左移，使 KI_3 分解，最终只能测得溶液中 I_2 和 KI_3 的总量。为此，向上述溶液中加入四氯化碳，然后充分振荡，使上述化学平衡和 I_2 在四氯化碳和水中的分配平衡同时建立（因 KI 和 KI_3 不溶于四氯化碳），则有

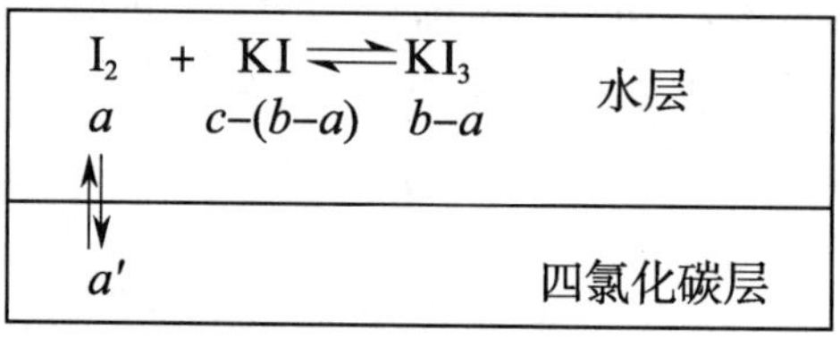

$I_2 \quad + \quad KI \rightleftharpoons KI_3$ $a \qquad c-(b-a) \qquad b-a$ $\updownarrow$	水层
a'	四氯化碳层

平衡后静置分层，分析四氯化碳层中 I_2 的浓度（设为 a'），由分配系数可以求出此时水层中 I_2 的浓度 $a\left(a=\frac{a'}{K}\right)$；而后取水层溶液进行分析，由 $Na_2S_2O_3$ 标准溶液滴定求出水层中 I_2 和 KI_3 的总浓度 b，则水层中 KI_3 的浓度为 $(b-a)$。设水层中 KI 的初始浓度为 c，平衡时 $[KI]=c-(b-a)$，至此可求得上述反应的平衡常数，即为

$$K_c=\frac{[KI_3]}{[KI][I_2]}=\frac{[b-a]}{a[c-(b-a)]} \qquad \text{式(2-6-3)}$$

（三）仪器和试剂

1. 仪器　恒温槽，25ml 和 100ml 量筒，25ml 和 5ml 滴定管，洗耳球，25ml 和 5ml 移液管，250ml 锥形瓶和 250ml 磨口锥形瓶。

2. 试剂 CCl_4,0.020 0mol/L I_2 的 CCl_4 溶液,0.02% I_2 的水溶液,0.100 0mol/L KI 水溶液,2.4% KI,0.020 0mol/L $Na_2S_2O_3$ 标准溶液,0.5%淀粉溶液。

(四)实验步骤

1. 控制恒温槽水温为25℃。

2. 按表2-6-1所列数据,将溶液配制于干燥的250ml磨口锥形瓶中。

3. 将上述锥形瓶置于恒温槽中,每隔10分钟取出振荡一次,约经1小时,或在室温下连续振荡约50分钟,静置,待系统两液层分层后,按表中所列数据取样进行分析。注意,为避免取四氯化碳层时水层进入移液管中,可用洗耳球使移液管尖端在不断鼓泡情况下通过水层。

4. 水层分析 先用 $Na_2S_2O_3$ 标准溶液滴定至淡黄色,加入2ml淀粉溶液指示剂,然后仔细滴至蓝色恰好消失。

5. CCl_4 层分析 在四氯化碳层样品中加入10ml水和2ml淀粉溶液,用 $Na_2S_2O_3$ 标准溶液滴定。滴定时要充分振荡,以使四氯化碳层中的 I_2 全部转入到水层(为增快 I_2 进入水层,可加入少量KI),细心滴至水层蓝色消失,四氯化碳层不再显红色即为终点。

6. 滴定结束后,所用四氯化碳及碘量瓶中剩余的溶液要倒入指定回收瓶中。

(五)数据记录和处理

1. 按表2-6-1记录数据

表2-6-1 溶液配制及实验数据

Table 2-6-1 Solution Preparation and Experimental Data

水浴温度(water bath temperature)________ KI浓度(concentration of KI)________

$Na_2S_2O_3$ 浓度(concentration of $Na_2S_2O_3$)________

		分配系数的测定(determination of partition coefficient)	化学平衡常数的测定(determination for equilibrium constant)	
No.		1	2	3
混合液的组成(composition of the liquid mixture)/ml	H_2O	200	50	0
	I_2 的 CCl_4 饱和溶液(saturated I_2 in CCl_4)	25	20	25
	KI	0	50	100
	CCl_4	0	5	0
分析取样量(sampling)/ml	CCl_4 层(CCl_4 layer)	5	5	5
	H_2O 层(water layer)	50	10	10
$V_{Na_2S_2O_3}$(ml)	CCl_4 层(CCl_4 layer)	1.		
		2.		
		3.		
		平均(average)		
	H_2O 层(water layer)	1.		
		2.		
		3.		
		平均(average)		
分配系数和平衡常数(partition coefficient and equilibrium constant)		K=	$K_{c,1}$ =	$K_{c,2}$ =
			平均(average)K_c =	

2. 计算 I_2 在四氯化碳层和水层的分配系数。

3. 计算反应 $KI+I_2 \rightleftharpoons KI_3$ 的平衡常数。

（六）思考题

1. 测定分配系数和平衡常数为什么要在恒温条件下进行?

2. 配制 1、2、3 号溶液进行实验的目的何在? 如何由滴定数据计算平衡时 KI、I_2 和 KI_3 的浓度(mol/L)?

3. 四氯化碳层滴定时,为什么要加入 KI? 加入 KI 对四氯化碳层中 I_2 的测定结果将产生何种影响?

（七）药学应用

分配系数可用于评价药物的亲脂性或亲水性的大小。油水分配系数对药物的吸收、在体内的生物利用度、药物与受体的作用以及药物代谢等密切相关。适宜的亲水性能够保证药物在体液中的溶解和转运,而一定的亲脂性能够保证药物进入细胞器或通过血脑屏障。在药物经皮吸收中,亲水性药物主要通过细胞间途径进入体内,而亲脂性药物则经由细胞内途径穿过角质层。

Experiment 6　Determination of Equilibrium Constant of Reaction and Partition Coefficient

1. Objective

1.1. To determine the equilibrium constant for reaction of iodine with iodide in water.

1.2. To determine the partition coefficient of iodine between water and carbon tetrachloride.

1.3. To learn the effect of temperature on equilibrium constant and partition coefficient.

1.4. To learn the pharmaceutical application of partition coefficient.

2. Principle

2.1. Partition coefficient

Iodine (I_2) can be distributed between two immiscible solvents of water and carbon tetrachloride(CCl_4) to form two separate layers in contact with each other. At a given temperature and pressure, the ratio of the equilibrium concentrations of the iodine in these two solvents will be constant.

$$I_2(\text{in } H_2O \text{ layer}) \rightleftharpoons I_2(\text{in } CCl_4 \text{ layer})$$

Thus we can have

$$K=\frac{c_{I_2}(\text{in } CCl_4 \text{ layer})}{c_{I_2}(\text{in } H_2O \text{ layer})} \tag{2-6-1}$$

where K is the partition coefficient for the system. The concentrations of the iodine in these two solvents can be determined by sodium thiosulfate($Na_2S_2O_3$) titration.

2.2. Equilibrium constant of reaction

I_2 is much more soluble in potassium iodide(KI) solution than in pure water because it can combine with KI to form tri-iodide ions(I_3^-). The reaction is as follows

$$I_2+KI \rightleftharpoons KI_3$$

At constant T and p, the equilibrium constant of the reaction can be expressed as

$$K_c = \frac{[KI_3]}{[I_2][KI]} \tag{2-6-2}$$

The equilibrium constant can be calculated in this form when concentrations of the substances at equilibrium are obtained. Usually I_2 in water is determined by $Na_2S_2O_3$ titration. But an equilibrium concentration of I_2 cannot be obtained if the titration is processed directly in the system because consumption of I_2 will result in the decomposition of KI_3 and what we determined is just a sum of c_{I_2} and c_{KI_3}. In this experiment an approach of solvent extraction with CCl_4 is used since I^- and I_3^- do not dissolve in CCl_4 but I_2 does.

After a partition equilibrium being established, the concentration of I_2 in aqueous layer can be calculated by partition coefficient as $a=\frac{a'}{K}$, provided the concentration of I_2 in CCl_4 is determined with a value of a'. The total concentration of I_2 and KI_3 in aqueous is determined by $Na_2S_2O_3$ titration with the value of b. Thus, the concentration of KI_3 in aqueous layer can be obtained by subtracting a from b. Assuming the initial concentration of KI is c, its equilibrium concentration should be $[KI]=c-(b-a)$. Therefore, the equilibrium constant can be calculated by the following equation

$$K_c = \frac{[KI_3]}{[KI][I_2]} = \frac{[b-a]}{a[c-(b-a)]} \tag{2-6-3}$$

3. Apparatus and Chemicals

3.1. Apparatus

Thermostat, 25ml and 100ml measuring cylinders, 25ml and 5ml burettes, suction bulb, 5ml and 25ml pipettes, 250ml cornical flasks with ground-in glass stopper, 250ml conical flasks.

3.2. Chemicals

CCl_4, 0.020 0mol/L I_2 in CCl_4, 0.02% I_2 in water, 0.100 0mol/L KI solution, 2.4% KI solution, 0.020 0mol/L $Na_2S_2O_3$ standard solution, 0.5% starch solution.

4. Procedures

4.1. Adjust the temperature of the thermostat to 25℃.

4.2. Prepare the test solution into 250ml cornical flask with ground-in glass stopper according to the data listed in table 2-6-1.

4.3. Place the flask into thermostat. Take it out and shake it every 10min for 1 hour, or keep shaking for 50min. Then let the bottle rest so that the layers separate. According to table 2-6-1, pipette out aqueous layer and organic layer into conical flasks respectively for iodine concentration analysis. To prevent from contamination by the aqueous layer when pipetting out the organic layer, use the suction bulb to continueously make the pipette tip bublling while inserting to the organic layer.

4.4. Aqueous layer analysis

Fill the burette with $Na_2S_2O_3$ standard solution. Titrate the aqueous layer to faint yellow, then add 2ml of starch solution. Continue to titrate with $Na_2S_2O_3$ standard solution till the blue color just disappears.

4.5. CCl_4 layer analysis

Add 10ml of water and 2ml of starch solution into sample of organic layer, titrate with

$Na_2S_2O_3$ standard solution. Keep shaking while titrating to completely transfer I_2 into water layer (for a quick transfer of I_2 into water layer, small amount of KI may be added). Titrate carefully till the blue color in water layer and the red color in CCl_4 layer just disappear.

4.6. Dispose of CCl_4 and solution in flask into the waste bottle.

5. Data recording and processing

5.1. Fill table 2-6-1 with the data measured and calculated.

5.2. Calculate the partition coefficient of iodine between water and CCl_4.

5.3. Calculate the equilibrium constant for reaction of $KI+I_2 \rightleftharpoons KI_3$.

6. Questions

6.1. What would be the effect on partition coefficient and equilibrium constant if the determination is not run under constant temperature?

6.2. Why three solutions of No.1 to No.3 are prepared in this experiment? How to calculate the equilibrium concentrations of KI, I_2 and KI_3 using the data of titration?

6.3. Why KI is allowed to add to the organic layer during the titration? What would be the effect on result if KI was added?

7. Pharmaceutical Applications

The partition coefficient represents the lipophilic/hydrophilic nature of the substance. Lipophilicity (or hydrophobicity) affects drug absorption, bioavailability, hydrophobic drug-receptor interactions, metabolism of molecules, as well as their toxicity. A certain degree of hydrophilic is required for drug dissolution and transportation in body fluid. A certain degree of lipophilicity is required to allow a drug to enter cellular organelles or to cross the blood-brain barrier. For percutaneous absorption of drugs through the skin, hydrophilic compounds may preferably partition into the intercellular, while the lipophilic compound may cross the stratum corneum through the intracellular route.

实验七　二组分部分互溶双液系统相图的绘制

（一）实验目的

1. 测定苯酚-水二组分部分互溶双液系统相互溶解度曲线，及系统的临界溶解温度与该温度下系统的组成。

2. 掌握液体相互溶解度的测定原理和方法。

3. 了解液体相互溶解度的概念。

4. 了解部分互溶双液系统的药学应用。

（二）实验原理

两种液体由于极性等性质有显著差别，在常温时只能有条件地相互溶解，即在某温度范围内，两种液体的相互溶解度都不大，只有当一种液体的量很少，另一种液体的量相对较多时，才能形成均匀的一相，超过一定温度范围或在其他配比下便会分层而呈两个液相平衡共存。例如，在常温下，将少量苯酚加到水中，它完全溶解而呈一相，是苯酚在水中的不饱和溶液。若继续加苯酚，会达到苯酚的溶解度，超过此极限，苯酚不再溶解在水里，这时出现两个液层：一层是苯酚在水中的饱和溶液，简称水层；另一层是水在苯酚中的饱和溶液，简称酚层。这两个液层称为共轭溶液,平衡共存。两液层的组成就是在该温度下,两组分的相互溶解度,即苯酚在水中的溶解度和水在苯酚中的溶解度。若继续增加苯酚的量,直至超过水在苯酚中的溶解度,又为单相,是水在苯酚中的不饱和溶液。图 2-7-1 是恒压下苯酚和水的温度-组成图。苯酚-水系统为具有最高会溶温度 T_c 的部分互溶双液系统,升高温度,其相互溶解度增大,达某温度以上时可以完全互溶。

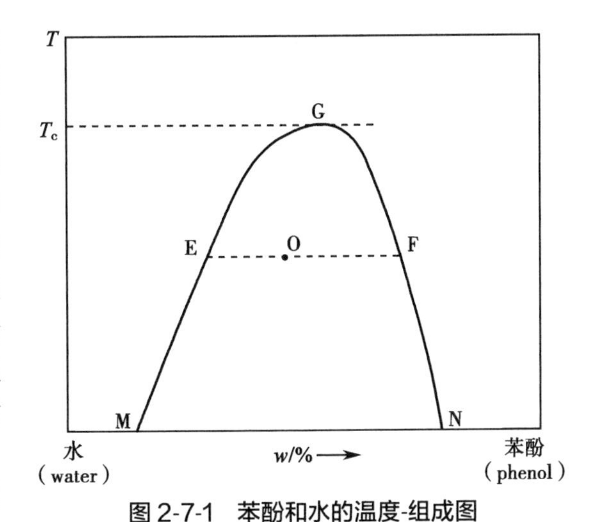

图 2-7-1 苯酚和水的温度-组成图

Fig. 2-7-1 Temperature-Composition Diagram for Phenol and Water System

液态部分互溶型的两组分相图可以通过测量二组分系统溶解度数据来绘制。本实验中,一系列不同组成的苯酚-水混合物被加热至完全互溶,然后冷却至溶液变浑浊并分成两个液层。以溶液开始分层的温度对组成作图,即可得到液态部分互溶型的两组分相图,从而可以确定临界溶解温度。

(三)仪器与药品

1. 仪器 磁力加热搅拌器 1 台,精密数字式温度计,分析天平,滴定管,容量约为 100ml 的大试管,500ml 烧杯,磁子,带孔胶塞,空气套管。

2. 药品 苯酚(分析纯),去离子水。

(四)实验步骤

1. 实验装置如图 2-7-2。

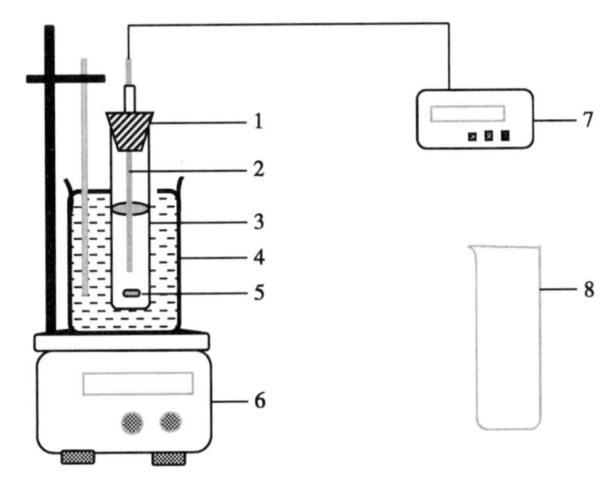

1. 胶塞(rubber plug);2. 测温探头(temperature measurement probe);3. 大试管(test tube);4. 烧杯(beaker);5. 磁子(stirring bar);6. 磁力加热搅拌器(magnetic stirrer with heating);7. ZT-2TB 精密数字式温度计(ZT-2TB precision digital thermometer);8. 空气套管(air casing)

图 2-7-2 相图测定装置图

Fig. 2-7-2 Sketch of Apparatus for Phase Diagram Determination

2. 精确称取5g苯酚于100ml干燥洁净的大试管中，用滴定管加入2.5ml水，注意保持管内混合物的液面低于水浴液面。

3. 接通加热电源，使外部水浴温度达到80℃左右，同时搅拌试管中的混合物，使两层充分混合，读取混合物由浑浊变为澄清时的温度，记作T_1。然后将试管从水中取出，擦干管外水分，放入空气套管内缓慢冷却，记录系统由澄清变为浑浊的温度T_2。T_1和T_2的差值不应超过0.2℃，否则须重复上述加热和冷却的操作，直到符合要求为止，其平均值作为混合物的溶解温度。

4. 在试管中分批加入去离子水，每次加0.5ml，共5次，以后每次加1ml。当温度超过一个最高值后，每次加2ml去离子水，共2次，以后加4ml，直到温度降到30℃以下。

5. 注意苯酚腐蚀性大，且易潮解，称量时应特别注意。

（五）数据处理

1. 列表记录实验数据。

2. 计算混合溶液中苯酚的质量分数。

3. 以温度为纵坐标，组成为横坐标，作水-苯酚系统的溶解度曲线。

4. 由溶解度曲线求最高临界溶解温度。

（六）思考题

1. 在实验操作中如何确保T_1和T_2的差值不超过0.2℃？

2. 在绘制的部分互溶双液系统相图中，应用相律说明帽形区内、外以及溶解度曲线上各点的自由度及意义。

（七）药学应用

在二组分部分互溶双液系统相图中，会溶温度的高低反映了一对液体间互溶能力的强弱，会溶温度越低，则两液体间相互溶解的能力越强，即溶解性越好。因此，可利用会溶温度的数据来选择优良的萃取剂。

Experiment 7 Drawing of Phase Diagram for a Partially Miscible Binary Liquid System

1. Objective

1.1. To construct the mutual solubility curve for phenol-water partially miscible liquid system and to determine the critical solution temperature.

1.2. To learn the principle and method for measuring miscibility of liquids.

1.3. To understand the miscibility of liquids.

1.4. To learn the pharmaceutical application of partially miscible binary liquid system.

2. Principle

Two liquids can only conditionally miscible with each other at a certain temperature, due to the differences in polarity and other properties. In another word, at a certain range of temperature, if the miscibility (mutual solubility) of these two liquids is low, they can only form a homogeneous phase when the quantity of one component is much less, and another is much large. Above a certain quantity range or in other proportion, two liquid layers form and both the solutions are in equilibrium. For example, at room temperature, a small amount of phenol will completely dissolve in water to form the unsaturated solution of phenol in water. If more phenol is gradually added to water, its solubility in water is reached. Then continuous addition of phenol will cause the separation of system to two liquid layers. The coexisting liquid phases are called conjugated solutions. They are saturated solutions of phenol in water and water in phenol, respectively. The compositions for the two layers are mutual solubilities of the two components under the temperature. If the amount of phenol to be added is over the solubility of water in phenol, then the two-phase system becomes one phase, which is the unsaturated solution of water in phenol. Figure 2-7-1 is the temperature-constitutional diagram for phenol and water under the constant pressure. Phenol-water system is partially miscible binary liquid system with the highest critical consolute temperature, the miscibility increases with increasing temperature. Above the critical temperature T_c, two components become completely miscible.

The phase diagram of partially miscible binary liquid system can be constructed by the mutual solubility data of two-component system. In this experiment, a series of different mixtures of phenol and water are prepared and heated to form a homogeneous solution. Then the mixture is cooled to a certain temperature at which the appearance of the system becomes turbidity and a second liquid appears. Plotting the temperate against compositions of the mixture gives the phase diagram of partially miscible binary liquid system.

3. Apparatus and chemicals

3.1. Apparatus

Magnetic stirrer with heating, precision digital thermometer, analytical balance, burette, tube has capacity of 100ml, 500ml beaker, stirring bar, perforated rubber plug and the air casing.

3.2. Chemicals

Phenol(AR), deionized water.

4. Procedures

4.1. Equipments are assembled according to Figure 2-7-2.

4.2. Add 5g of phenol into 100ml dry and clean tube, and then add 2.5ml of water using burette into the tube. Keep a lower surface of the liquid mixture in the test tube than that of the water bath.

4.3. Turn on the power supply. Adjust the temperature of external water bath to 80℃. Heat the solution, stirring continuously for a complete mixing of the mixture. Record the temperature at which turbidity just disappears. Note it as T_1. Take the tube out from the bath, dry the outside. Insert the tube into the air casing to allow the tube to cool slowly. Record the temperature T_2 at which the system changes from clear to cloudy. The difference between T_1 and T_2 should not be more than 0.2℃, otherwise the heating and cooling operations must be repeated until it meets the requirement. The average of T_1 and T_2 is recognized as the miscibility temperature.

4.4. Add another 0.5ml of water into the tube in turn and repeat the experiment five times. Then add 1ml of water each time and run the experiment until the temperature exceeds a peak. Add another 2ml of water two times, and then 4ml water each time till the temperature drops below 30℃.

4.5. Attention

Phenol is corrosive and is easily deliquesce, special attention should be paid to weighing it.

5. Data recording and processing

5.1. Record the experiment data in list.

5.2. Calculate the mass fraction of phenol in the mixture.

5.3. Solubility curve of phenol water is constructed with temperature as a vertical axis, composition as a horizontal axis. .

5.4. Determine the maximal critical solution temperature.

6. Questions

6.1. How to ensure difference between T_1 and T_2 is less than 0.2℃ in the experiment?

6.2. Indicate the degree of freedom, significance of each point inside or outside the cap area, as well as on the solubility curve.

7. Pharmaceutical Applications

In phase diagram of partially miscible binary liquid system, the consolute temperature reflects the miscibility of two liquids. Lower consolute temperature means higher miscibility of two liquids, namely the better solubility. Therefore, we can use consolute temperature data to choose excellent extraction agent.

实验八 完全互溶双液系统平衡相图的绘制

(一) 实验目的

1. 用回流冷凝法测定不同浓度环己烷-乙醇系统的沸点和气、液两相平衡组成,绘制沸点-组成图。

2. 正确掌握阿贝折光仪的使用方法。

3. 通过实验进一步理解相图和相律的基本概念及分馏原理。

4. 了解相图的药学应用。

(二) 实验原理

任意两个在常温时为液态的物质混合组成的系统称为双液系统。两种液体若能按任意比例溶解,称为完全互溶双液系统。恒压下完全互溶双液系统的沸点-组成图,揭示了气液两相平衡时,沸点和两相成分间的关系。完全互溶双液系统的沸点-组成图有三种类型:

(1) 溶液沸点介于两种纯组分沸点之间,如苯与甲苯的双液系统,见图 2-8-1(a)所示。

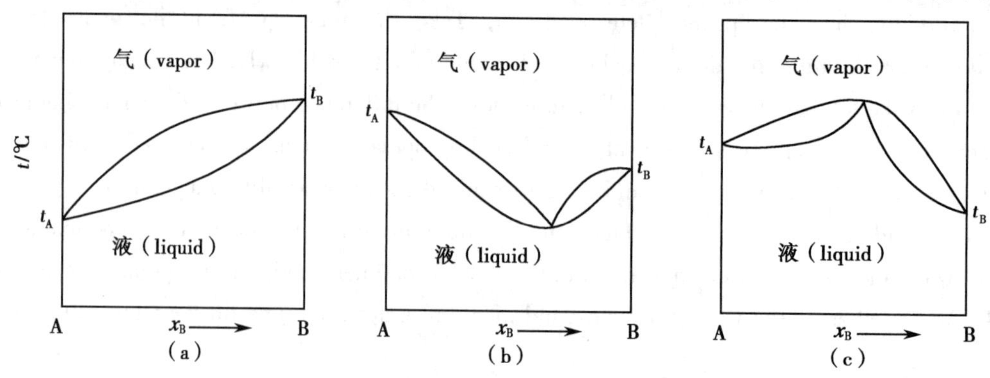

图 2-8-1 完全互溶双液系统的沸点组成图

Fig. 2-8-1 Temperature-Composition Diagram for Binary Systems

这类双液系统可用分馏法从溶液中分离出两个纯组分。

(2) 溶液有最低恒沸点,如环己烷-乙醇、水-乙醇的双液系统,见图 2-8-1(b)所示。

(3) 溶液有最高恒沸点,如硝酸-水的双液系统,见图 2-8-1(c)所示。

具有最低或最高恒沸点的双液系统,在恒沸点处液相和气相组成相同,这时的混合物称为恒沸混合物。恒沸物不能用简单分馏法分离出两个纯组分,需采用特殊分馏方法。

本实验中的环己烷-乙醇双液系统具有最低恒沸点,在常压下对不同组成的样品进行回流冷凝达到平衡,测定气液两相平衡时的沸点和两相组成,绘制沸点-组成图。

测定沸点和平衡组成的装置称为沸点测定仪(图 2-8-2),主要部件为带有回流冷凝管的沸点仪,里面装有加热棒和温度传感器。冷凝管底部有半球形小室,用以收集冷凝下来的气相样品。气相与液相平衡时的沸点可由沸点测定仪的控制面板直接读出,两相的组成可通过测定折光率对照标准曲线计算得到。

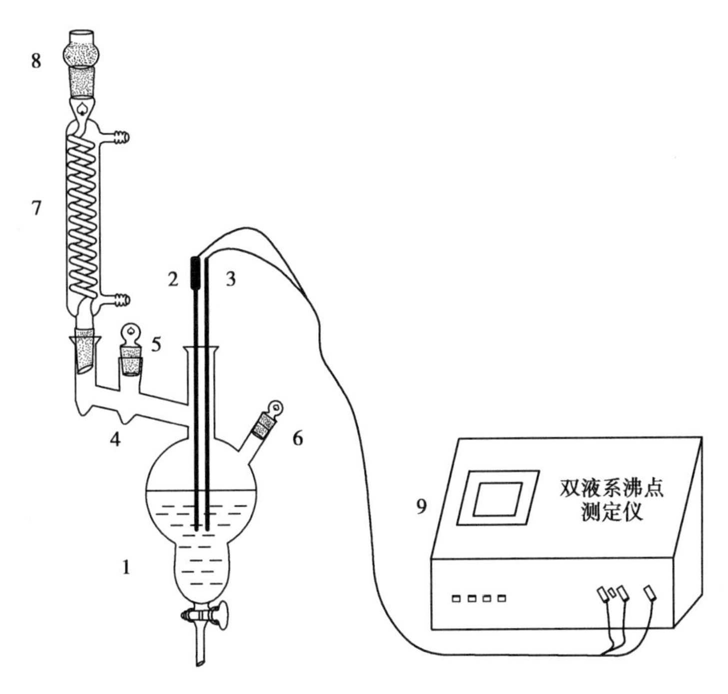

1. 沸点仪(boiling apparatus);2. 加热棒(heating rod);3. 温度传感器(thermosensor);4. 气相冷凝液接收管(condensate receiving vessel);5. 气相冷凝液取样口(sampling orifice);6. 加液口(feeding entrance);7. 冷凝管(condenser);8. 干燥管(drying tube);9. 沸点测定仪(包括调压器)(boiling point tester, including voltage regulator)

图 2-8-2 沸点测定装置

Fig. 2-8-2 Device for Boiling Point Determination

(三)仪器和试剂

1. 仪器 沸点仪,沸点测定仪(调压器),阿贝折光仪(接超级恒温槽),硅胶干燥管,温度传感器,吸液管,样品管。

2. 试剂 环己烷(分析纯),无水乙醇(分析纯),不同浓度的环己烷-乙醇混合样品 8 份。

(四)实验步骤

1. 标准曲线的绘制 配制环己烷体积分数为 0、0.1、0.2、0.3、0.4、0.5、0.6、0.7、0.8、0.9 和 1.0 的环己烷-乙醇溶液,在 25℃时测定折光率,作折光率对浓度的标准曲线。标准曲线也可用 Excel 软件中的二次三项式拟合得到,方程式为

$$y=an^2+bn+c \qquad 式(2\text{-}8\text{-}1)$$

式中 n 为折光率;y 为组成(体积分数);a、b、c 为拟合参数。拟合相关系数 r 应大于 0.999 9。

2. 折光仪的使用训练 连接折光仪的超级恒温槽事先调至 25℃。以纯乙醇(或纯环己烷)为样品,练习折光率测定的操作,直到能对样品进行迅速准确的测定。每次测定后,应对折光仪中的样品池进行清洁干燥处理。

3. 沸点测定仪的安装 将干燥的沸点仪按图 2-8-2 安装好,塞紧带有温度传感器和加热棒的塞子,冷凝管上部装好硅胶干燥管。

4. 沸点测定 将待测样品约 30ml 从加液口倒入沸点仪中(注意加热棒和温度传感器应浸入溶液中),打开冷凝水,接通电源,调整输出电压约 20~30V(勿使电压过大,以免发生事故),使液体缓缓加热升温至沸腾。液体沸腾后,保持回流数分钟,并将接收管中的最初气相冷凝液倒回到液相中 2~3 次,在气液充分平衡后(此时温度恒定),读取沸点,并停止加热。

5. 组成测定 用干燥的吸管分别吸取气相冷凝液(气相样品)和残馏液(液相样品),用阿贝折光仪迅速测定其折光率。测定完毕,将原溶液全部由下口放出,收集在回收瓶中。沸点仪、取样吸管用电吹风作干燥处理。

6. 重复步骤 4、5,同法测定其他样品的沸点和气、液两相的折光率。

7. 准确读取实验时的大气压。

8. 实验结束,关闭电源,关闭冷凝水,整理实验台。

9. 实验注意事项

(1) 为保证系统内外气压的一致,系统不能密闭。

(2) 测定样品的折光率时,先测定气相冷凝液,再测定液相冷凝液。

(3) 加热棒应浸入溶液 2cm 左右。

(五)数据记录与处理

1. 实验数据记录(表 2-8-1)

表 2-8-1 样品的折光率和组成

Table 2-8-1 Refractive Indices and Compositions of the Samples

室温(room temperature)________℃ 大气压(atmospheric pressure):________ kPa

折光率测定温度(temperature for refractive determination):25℃

样品序号 No.	沸点(boiling point)/℃	气相(gas phase)		液相(liquid phase)	
		折光率(refractive)	环己烷/V%(cyclohexane/V%)	折光率(refractive)	环己烷/V%(cyclohexane/V%)
1					
2					
…					
9					
10					

按标准曲线方程，由折光率数据计算相应的组成。

1号和10号样品为纯乙醇和纯环己烷，可不作实验测定。其沸点根据当天的大气压按克劳修斯-克拉珀龙方程计算：

$$\ln \frac{p_2}{p_1} = -\frac{\Delta H_{vap}}{R}\left(\frac{1}{T_2} - \frac{1}{T_1}\right) \qquad 式(2\text{-}8\text{-}2)$$

乙醇的正常沸点为351.7K，摩尔气化热 ΔH_{vap} 为39.380kJ/mol。环己烷的正常沸点为353.9K，摩尔气化热 ΔH_{vap} 为29.952kJ/mol。

2. 根据实验数据绘制环己烷-乙醇溶液的沸点-组成相图。

（六）思考题

1. 将含环己烷30%的环己烷-乙醇溶液在101.325kPa进行精馏时，如塔效率足够高，可以得到什么馏出液和残馏液？

2. 将恒沸混合物进行精馏时可以得到什么结果？

3. 如何判断气液两相已达平衡？

4. 本实验的误差来源有哪些？

（七）药学应用

相图的绘制在药学中有着重要的应用。例如：在药物制剂中，微乳相图的绘制指导处方的设计；在药物合成中可气化物质的分离提纯，如蒸馏、分馏和精馏，溶剂的回收；在药物分析中气相色谱的分离原理的建立也是借助该相图完成的。绘制相图可以帮助我们了解最低恒沸点、不同温度压力下物质的状态等信息，对指导药物制剂的设计有重要意义。

（八）阿贝折光仪

折光率是很多液体药物规定的理化常数指标之一，测定折光率可以鉴别药液的纯度或测出其含量。

1. 阿贝折光仪的原理　阿贝折光仪是根据临界折射现象设计的，如图2-8-3所示。试样置于棱镜的界面上，而棱镜的折射率 n_p 大于试样的折射率 n，如果入射线1正好沿着棱镜与试样的界面入射，其折射光则为1′，此时入射角 $i_1=90°$，折射角为 γ_c，此即称为临界折射角。大于临界角的区域为暗区，小于临界角的区域为亮区。已知在温度、压力不变的条件下，入射角 i_m 和折射角 γ_M 与两种介质的折光率 n（介质M的）和 N（介质m的）的关系为 $\frac{\sin i_m}{\sin\gamma_M} = \frac{n}{N}$，由此可知

$$n = n_p \frac{\sin\gamma_c}{\sin 90°} = n_p \sin\gamma_c \qquad 式(2\text{-}8\text{-}3)$$

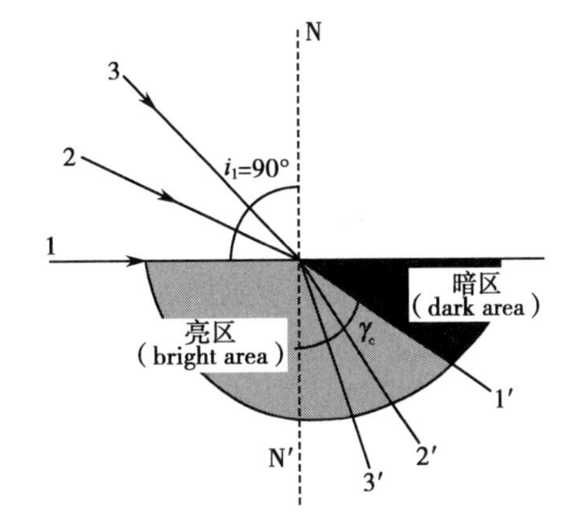

图2-8-3　明暗分界线的形成
Fig.2-8-3　Formation of Shade Dividing Line

显然，如果已知棱镜的折射率 n_p，并在温度、单色光波长都保持恒定值的实验条件下，测定临界角 γ_c，就能计算出试样的折光率 n。

根据测定临界折射角确定折光率的原理，设计成测定折光率的仪器，最常用的是阿贝折光仪。

2. 阿贝折光仪的结构 如图 2-8-4 所示，是一种典型的阿贝折光仪。其中心部件是由两块直角棱镜组成的棱镜组，下面一块是可以启闭的辅助棱镜，其斜面是磨砂的毛玻璃面，液体试样夹在辅助棱镜与测定棱镜之间展开为一薄层。光线由光源经反射镜反射至辅助棱镜，并在磨砂面上产生漫反射，因此，各个方面都有从试样层进入测量棱镜的光线，从测量棱镜的直角边上方，可观察到临界折射现象。由于液体折射率与所用的光线波长和温度有关，在折光仪的上下棱镜的外面设有恒温水接头，以保持棱镜恒温。

转动棱镜组的转轴可以调整棱镜的角度，使临界线正好落在测量望远镜视野的十字形准线交点上，由于刻度盘与棱镜组是同轴同步转动的，因此，从读数放大镜的标尺中就可读得该液体的折射率数值。

刻度盘上有两行示值，右边的一行是在以日光为光源的条件下，直接换算成相当于钠光 D 线的折光率 n_{D}（从 1.300 0~1.700 0），左边的一行，示值为 0~95%，是工业上用折光仪测量固体物质在水溶液中的浓度的标尺。

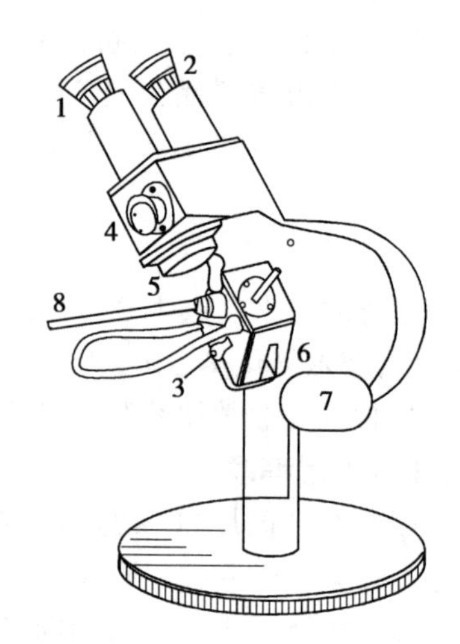

1. 目镜（ocular lens）；2. 读数放大镜（reading glass）；3. 恒温水接头（homoeothermic connecter）；4. 消色补偿器（chromaticity complementer）；5、6. 棱镜（prisms）；7. 反射镜（reflector）；8. 温度计（thermometer）

图 2-8-4 阿贝折光仪
Fig. 2-8-4 Abbe Refractometer

3. 液体折光率的测定

（1）仪器的安置：将折光仪置于靠窗的桌上或普通白炽灯前，但勿暴露于直照的日光中，以免试样迅速挥发。安上专用温度计，用橡皮管将测定棱镜和辅助棱镜上保温夹套的进、出水口与超级恒温槽的进、出水口串联接通，恒温温度以折光仪的温度计读数为准，一般选控为 20.0℃±0.1℃ 或 25.0℃± 0.1℃。

（2）加试样：松开锁钮，开启辅助棱镜，使其磨砂面约处于水平位置，用滴管加少量丙酮清洗镜面，使难挥发的污物逸走（注意：滴管尖勿碰触镜面！），必要时可用拭镜纸轻拭镜面（注意：勿用滤纸擦拭！），待镜面干燥后，加试样 1~3 滴于辅助棱镜的磨砂面上，迅速闭合辅助棱镜，适当旋紧锁钮（若试样易挥发，可从闭合后两棱镜侧面的加液槽加入）。

（3）对光：转动测量手柄，使刻度盘标尺的示值为最小（1.300 0），调节反射镜使光线进入棱镜组，同时从测量望远镜中观察，使视场最亮为止。调节望远镜上部的目镜焦距，使视场的准线最清晰。

（4）粗调：转动测量手柄，使刻度盘标尺的示值逐渐增大，直至观察到视场中出现彩色光带或黑色临界线为止。

（5）消色散：转动消色散手柄，使彩色消失而呈现一清晰的黑白临界线。

（6）精调：再转动测量手柄，使临界线移动到十字形准线的交叉点上。如有彩色产生，须再调消色散手柄消色，使临界线黑白清晰。

（7）读数：将刻度盘罩壳侧面上方的镀铬小窗打开至恰当位置，将光线反射入罩内，然后从望远镜中读出相应的示值。为了减少偶然误差，应转动手柄，重复调节测定三次，相互间相差不能大于 0.002，取平均值。

（8）仪器的校正：折光仪刻度盘上标尺的零点，有时会产生移动，须加以校正。校正的方法是用一已知折光率的标准液体，一般用纯水按上述方法进行测定，将测定平均值与标准值比较，其差值即为校正值。纯水的 $n_D^{20}=1.332\ 5$，在 15～30℃时的温度系数为-0.000 1/℃。

Experiment 8　Drawing of Phase Diagram for a Miscible Binary Liquid System

1. Objective

1.1. To determine the boiling point and gas-liquid two-phase equilibrium composition of cyclohexane-ethanol systems with different concentrations by reflux condensation method and draw the boiling point-composition diagram.

1.2. To learn the principle and operation of Abbe refractometer.

1.3. To grasp the basic concept of phase diagram, phase rule and the theory of fraction distillation.

1.4. To learn the pharmaceutical applications of phase diagram.

2. Principle

When a system consists of two kinds of liquid at room temperature is called a binary liquid system. If a system consists of two liquids which are soluble (miscible) in one another in all proportions, then the system is called miscible binary liquid system. For such a system, temperature-composition phase diagram can be used to express the relationship between the boiling point and composition when liquid-vapor equilibrium is reached. Usually, there are three types of temperature-composition phase diagrams for the miscible binary liquid systems:

(1) The boiling point of the solution lies between that of two pure liquids, such as the binary liquid system of benzene and toluene, as shown in figure 2-8-1(a). In this kind of binary liquid system, two pure components can be separated from the solution by fractionation.

(2) System with a minimum azeotropic point in the temperature-composition phase diagram is shown in Fig. 2-8-1(b), such as water-alcohol, and cyclohexane-alcohol liquid-liquid system.

(3) System with a maximum azeotropic point in the temperature-composition phase diagram is shown in Fig. 2-8-1(c), such as nitric acid-water liquid-liquid system.

For system with high-boiling or low-boiling azeotropic point, the liquid and vapor composition is the same and this mixture is referred to as azeotropic mixture. Special fractional distillation method is required for the separation of constant boiling mixture since simple fractional distillation method doesn't work here.

In this experiment, the cyclohexane-ethanol binary liquid system has the lowest azeotropic point. The samples with different compositions are refluxed and condensed at atmospheric pressure to achieve equilibrium. The boiling point and the composition of the gas-liquid binary phase equilibrium are determined, and the *T-x* diagram is drawn.

The device for determining the boiling point and equilibrium composition is called the boiling point tester (Fig. 2-8-2). The main component is a boiling point apparatus with reflux condensing tube, and a heating rod and a temperature sensor are equipped. There is a hemispherical chamber at the bottom of the condensing tube to collect condensed gas samples. The boiling point of the equilibrium between the gas phase and the liquid phase can be read directly from the control panel of the boiling apparatus, and the composition of the two phases can be calculated by measuring the refractive index compared with the standard curve.

3. Apparatus and chemicals

3.1. Apparatus

Boiling apparatus, boiling point tester (including voltage regulator), Abbe refractometer (connected to super thermostatic tank), drying tube with silica gel, temperature sensor, pipettes, sample holders.

3.2. Chemicals

Cyclohexane (AR), ethanol (AR), eight cyclohexane-ethanol solutions of different compositions.

4. Procedures

4.1. Construction of the calibration curve

A series of cyclohexane-ethanol solutions containing 0.1, 0.2, 0.3, 0.4, 0.5, 0.6, 0.7, 0.8 and 0.9 volume fractions of cyclohexane are prepared. The refractive index is measured at 25℃ and the calibration curve of refractive index is drawn.

The calibration curve can also be obtained by quadratic trinomial function fitting using Excel software with the equation as follows

$$y = an^2 + bn + c \tag{2-8-1}$$

where n is the refractive index, y is the composition, a, b and c are fitting coefficients. The correlation coefficient r of curve fitting should be larger than 0.999 9.

4.2. Operation practice of Abbe refractometer

Adjust the super thermostatic bath to 25℃. Choose pure ethanol (or pure cyclohexane) as model sample to practice the operation of Abbe refractometer until a quick and accurate measurement is obtained. Clean and dry the sample cell of the refractometer after each measurement.

4.3. Assembly of the boiling apparatus

Assemble the apparatus as shown in Fig. 2-8-2, place the heating wire close to the center of the flask at the bottom, stopple the flask, taking care not to contact the mercury bulb with heating coil, and install a silica gel drying tube to the upper part of the condenser.

4.4. Determination of boiling point

Add about 30ml of the sample into the flask, make sure that the heating rod and temperature sensor are immersed fully in the solution. Turn on the cooling water and power supply. Adjust the output voltage of the electric transformer to about 20~30V (make sure the voltage isn't too large, so as to avoid accidents), to heat slowly until the solution boils. Then keep a vigorous boiling for several minutes. Pour back the initial gas condensate in the receiving tube into the liquid phase for 2 or 3 times. After the gas-liquid balance is fully established and the thermometer reading has become constant, record the temperature and stopped the heating.

4.5. Measurement of the composition

Use dry pipette to take samples of the gas phase condensate (gas phase sample) and the residual distillate (liquid phase sample), and determine the refractive quickly by Abbe refractometer. After the determination, release all the original solution from the lower orifice and collect in the recovery bottle. Dry the boiling point meter and sampling pipette with a blow dryer.

4.6. Repeat procedures 4.4 and 4.5 and determine the boiling point and refractive indices of other samples.

4.7. Read the atmospheric pressure at the experiment.

4.8. Turn off both the power supply and cooling water. And clear up the experiment table.

4.9. Attentions

4.9.1. The system shouldn't be airtight to ensure the consistency of pressure inside and outside.

4.9.2. Determine the refractive index of gas condensate before that of liquid condensate.

4.9.3. The heating rod should be immersed in the solution 2cm or so.

5. Data recording and processing

5.1. Fill table 2-8-1 with the data measured and calculated. Calculte the relevant composition according to the refractive index based on the standard curve equation.

Samples 1 and 10 are pure ethanol and pure cyclohexane, which do not have to be determined experimentally. Its boiling point is calculated according to the Clausius-Clayperon equation according to the atmospheric pressure of the day:

$$\ln\frac{p_2}{p_1}=-\frac{\Delta H_{\mathrm{vap}}}{R}\left(\frac{1}{T_2}-\frac{1}{T_1}\right) \tag{2-8-2}$$

The normal boiling point of ethanol is 351.7K and the molar heat of gasification ΔH_{vap} is 39.380kJ/mol. The normal boiling point of cyclohexane is 353.9K and the mole heat of gasification ΔH_{vap} is 29.952kJ/mol.

5.2. According to the experimental data, draw the temperature-composition phase diagram of binary liquid system of cyclohexane and ethanol.

6. Questions

6.1. If a mixture of 30% cyclohexane solution is distilled at 101.325kPa, what will be the distillate and residue?

6.2. When azeotropic mixture is treated by fraction distillation, what product can be obtained?

6.3. How to judge the equilibrium for gas phase and liquid phase?

6.4. Discuss the main source of errors in the experiment.

7. Pharmaceutical application

Phase diagram plays an important role in pharmacy. For example, in pharmaceutic preparation, the phase diagram of microemulsion guides the formulation design. In drug synthesis, it can be used in distillation, fractionation and rectification of gasifiable substances and solvent recycle in drug synthesis. In drug analysis, it is also needed for the establishment of the separation principle in gas chromatography. It helps us to understand the minimum azeotropic point and to know information about the states of matter under different temperature and pressure. Moreover, it is essential to guide pharmaceutical preparations.

8. Abbe refractometer

The refractive index is one of the physical and chemical constants specified by many liquid drugs. The purity or content of the liquid can be identified by measuring the refractive index.

8.1. The principle of Abbe refractometer for measuring the refractive index of liquid media

The Abbe refractometer is designed according to the critical refraction phenomenon, as shown in figure 2-8-3. The sample is placed on the interface of the prism, and the refractive index n_p of the prism is larger than the refractive index n of the sample. If the incident ray 1 happens to be incident along the interface between the prism and the sample, the refracted light is 1', and the incident angle is $i_1 = 90°$, and the refractive angle is γ_c, which is called the critical refractive angle. The area larger than the critical angle is the dark area, and the area less than the critical angle is the bright area. It is known that under the condition of constant temperature and pressure, the incident angle i_m and refractive angle γ_M can be related to the refractive index n(medium M) and N(medium m) of the two media: $\frac{\sin i_m}{\sin \gamma_M} = \frac{n}{N}$, and then

$$n = n_p \frac{\sin \gamma_c}{\sin 90°} = n_p \sin \gamma_c \tag{2-8-3}$$

Obviously, if the refractive index n_p of the prism is known, and under the experimental condition that the temperature and the wavelength of monochromatic light are constant, the refractive index n of the sample can be calculated by measuring the critical angle γ_c.

Instrument for measuring refractive index based upon the principle of determining the critical angle of a substance as a measure of refractive index is called refractometer. The most common type of refractometer is Abbe refractometer.

8.2. Structure of Abbe refractometer

Fig.2-8-4 shows a typical Abbe refractometer. The central part is a prism group composed of two right-angle prisms, the lower one is an auxiliary prism that can be opened and closed, its slope is a frosted ground glass surface, and the liquid sample is sandwiched between the auxiliary prism and the measuring prism into a thin layer. The light is reflected from the light source to the auxiliary prism through the mirror and produces diffuse reflection on the frosted surface. Therefore, there is light from the sample layer into the measuring prism in all aspects, and the critical refraction can be observed from above the right-angled edge of the measuring prism. As the refractive index of the liquid is related to the wavelength and temperature of the light used, a constant temperature water connector is arranged outside the upper and lower prism of the refractometer to keep the prism constant temperature.

The rotation axis of the prism group can adjust the angle of the prism so that the critical line falls on the intersection of the cross line of the field of view of the telescope. Because the dial rotates synchronously with the prism group, the refractive index of the liquid can be read from the ruler of the reading magnifier.

There are two lines of values on the dial, the right one is directly converted to the refractive index n_p(from 1.300 0 to 1.700 0) equivalent to the sodium D line under the condition of sunlight as the light source, and the left row, with an indication of 0 ~ 95%, is an industrial scale for measuring the concentration of solid substances in aqueous solution with a refractometer.

8.3. Determination of refractive index of liquids

8.3.1. Construction of the instrument: Place the refractometer on the table near the window or in front of the ordinary incandescent lamp, but do not expose to direct sunlight to avoid rapid volatilization of the sample. A special thermometer is installed, and the inlet and outlet of the thermal insulation jacket on the measuring prism and the auxiliary prism are connected in series with the inlet and outlet of the super constant temperature trough, and the constant temperature is based on the thermometer reading of the refractometer. The general selection and control is 20.0℃ ±0.1℃ or 25.0℃ ±0.1℃.

8.3.2. Sample addition: Release the lock button, open the auxiliary prism, make the frosted surface about horizontal, clean the surface with a few drops of acetone to make the non-volatile dirt escape (note: not to touch the surface with the pipette to avoid scratching the refracting prism!) If necessary, gently wipe the prism surface with wiping paper (note: do not wipe with filter paper!) After the surface is dry, add 1 to 3 drops of the sample to the prism surface, quickly close the auxiliary prism and tighten the locking knob properly (if the sample is volatile, it can be added from the liquid trough on the side of the closed prism).

8.3.3. Light focusing: Turn the measuring handle so that the indication value of the dial scale is minimum (1.300 0), adjust the reflector so that the light enters the prism group, and observe from the measuring telescope so that the field of view is the brightest. Adjust the focal length of the eyepiece at the top of the telescope to make the alignment of the field of view the clearest.

8.3.4. Coarse adjustment: Turn the measuring handle to gradually increase the indicating value of the dial scale until a colored light band or black boundary is observed in the field of view.

8.3.5. Dispersion elimination: Turn the achromatic handle to make the color disappear and show a clear black-and-white boundary.

8.3.6. Precise adjustment: Turn the measuring handle again to move the critical line to the intersection of the cross line. If there is a color, it is necessary to adjust the achromatic handle to make the borderline black and white clear.

8.3.7. Reading: Open the chrome-plated window above the side of the dial housing to the proper position, reflect the light into the cover, and then read the corresponding value from the telescope. In order to reduce the accidental error, the handle should be turned and the measurement should be repeated for three times. The difference between them should not be more than 0.002, and the average value should be taken.

8.3.8. Adjustment of the instrument: The zero point of the ruler on the dial of the refractometer sometimes shifts and needs to be corrected. The method of correction is to use a standard liquid with a known refractive index, which is generally determined by pure water according to the above method. Comparing the measured average value with the standard value, the difference is the correction value. For pure water, $n_D^{20} = 1.332\ 5$, the temperature coefficient is $-0.000\ 1/℃$ at 15~30℃.

实验九 二组分简单低共熔系统相图的绘制

（一）实验目的

1. 应用步冷曲线的方法绘制 Cd-Bi 二组分系统的相图。
2. 掌握热电偶温度计的基本原理和使用方法。
3. 了解低共熔相图的药学应用。

（二）实验原理

二组分相图的应用非常广泛，如钢铁冶炼、石油化工、陶瓷技术，产品分离等。绘制相图的方法也很多，其中较常用的方法是热分析法。在一定压力下把系统从高温逐渐冷却，作温度对时间变化曲线，即步冷曲线。系统若有相变，必然伴随有热效应，即在其步冷曲线中会出现转折点。从步冷曲线有无转折点就可以知道有无相变。测定一系列组成不同样品的步冷曲线，从步冷曲线上找出各相应系统发生相变的温度，就可绘制出被测系统的相图，如图 2-9-1 所示。

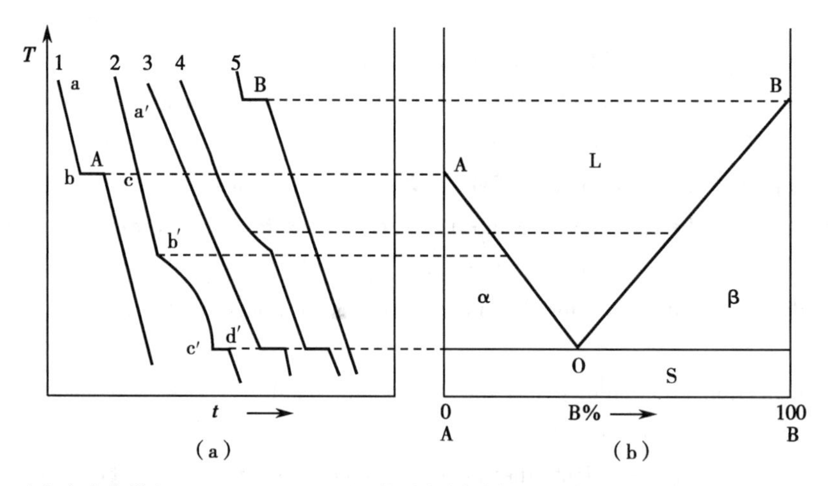

(a)步冷曲线(Step Cooling Curves);(b)固液相图(Solid-Liquid Phase Diagram)

图 2-9-1 步冷曲线绘制固液相图

Fig. 2-9-1 Solid-Liquid Phase Diagram Plotting by Step Cooling Curves

纯物质的步冷曲线如图 2-9-1(a)中线 1、线 5 所示，从高温冷却，开始降温很快，ab 线的斜率决定于系统的散热程度。冷到 A 的熔点时，固体 A 开始析出，系统出现两相平衡（溶液和固体 A），此时温度维持不变，步冷曲线出现 bc 的水平段，直到其中液相全部消失，固体温度下降。

混合物步冷曲线（如线 2、线 4）与纯物质的步冷曲线（如线 1、线 5）不同。如线 2 起始温度下降很快（如 a′b′段），冷却到 b′点的温度时，开始有固体析出，这时系统呈两相，因为液相的成分不断改变，所以其平衡温度也不断改变。由于凝固热的不断放出，其温度下降较慢，曲线的斜率较小（b′c′段）。到了低共熔点温度后，系统出现三相，温度不再改变，步冷曲线又出现水平段 c′d′，直到液相完全凝固后，固体温度又迅速下降。

曲线 3 表示其组成恰为最低共熔混合物的步冷曲线，其图形与纯物质相似，但它的水平段是三相平衡。

用步冷曲线绘制相图是以横轴表示混合物的成分，在对应的纵轴标出开始出现相变(即步冷曲线上的转折点)的温度，把这些点连接起来即得相图。

图 2-9-1(b)是一种形成简单低共熔混合物的二组分系统相图。图中 L 为液相区；左三角区为纯 A 和液相共存的二相区，右三角区为纯 B 和液相共存的二相区，水平线段表示 A、B 和液相共存的三相共存线(不包括线与纵轴相交的两端点)；水平线段以下表示纯 A 和纯 B 共存的二相区；O 为低共熔点。

(三) 仪器和试剂

1. 仪器 500W 加热电炉，热电偶，陶质坩埚，电位计，硬质试管，变压器。

2. 试剂 镉(化学纯)，铋(化学纯)，液状石蜡等。

(四) 实验步骤

1. 配制不同质量百分比的铋、镉混合物各 50g(镉含量分别为 0、15%、25%、40%、55%、75%、90%、100%)，分别放在 8 只硬质试管中，再各加入少许液状石蜡(约 3g)，以防止金属在加热过程中接触空气而氧化。

2. 依次测定纯镉、纯铋和含镉质量百分比为 90%、75%、55%、40%、25%、15%样品的步冷曲线。将样品管放在加热电炉中，控制电压在 160V 左右，让样品熔化，同时将热电偶的热端(连玻璃套管)插入样品管中，冷端插入冰水中，待样品熔化后，停止加热。用热电偶玻璃套管轻轻搅拌样品，使各处温度均匀一致，避免发生过冷现象。当样品均匀冷却时，每隔 1 分钟测定电动势一次，直到步冷曲线的水平部分以下为止。用纯镉和纯铋来校正热电偶，已知纯镉和纯铋的熔点分别为 594.3K 和 544.6K。

3. 注意事项

(1) 样品冷却过程中，冷却速度保持在 6~8K/min 之间(当环境温度较低时，可加一定的低电压于电炉中)，热电偶的热端放在样品中央，离样品管管底不小于 1cm，否则将受外界的影响，而不能真实反映被测体系的温度。

(2) 小心操作自制样品管，防止烫伤。

(3) 样品混合要求均匀。

(4) 控制冷却速度，不要太快，防止拐点和平台不明显。

(五) 数据处理

1. 利用所得步冷曲线，绘制 Cd-Bi 二组分体系的相图，并指出相图中各区域的相平衡。

2. 从相图中求出低共熔点的温度及低共熔混合物的成分。

(六) 思考题

1. 对于不同成分混合物的步冷曲线，其水平段有什么不同？

2. 用加热曲线是否可作相图？

3. 作相图还有哪些方法？

(七) 药学应用

固体分散体是提高药物吸收效果和生物利用度的有效方法，利用低共熔相图原理，使药物与载体以低共熔比例共存时，制成的药物具有均匀的微细分散结构，增加了药物的表面积，可大大改善其溶出速度。如伊曲康唑和泊洛沙姆 188 制成的低共熔混合物的溶出速率是纯伊曲康唑和药物与载体的物理混合物的 4.66 倍和 2.96 倍；48%尿素与 52%磺胺噻唑制成的低共熔混合物的溶出速度是纯磺胺噻唑的 12 倍。

(八) 其他测量方法

1. 为了控制升温和冷却速度，便于学生观察温度变化，可以采用 SWKY-1 型数字控温仪和 KWL-09 型可控升降温电炉进行实验。

2. 原有实验的样品管都为自制而且每次只能测定一个样品，为了减轻学生工作量可以使用多通道金属相图测定装置（如 JX-3D8 型）。此装置可以一次测量 8 个样品，而且样品已经按要求封装完好，可以多次使用。利用仪器所带软件可以将实验所得拐点和平台输入得到二元金属相图，装置如图 2-9-2 所示。

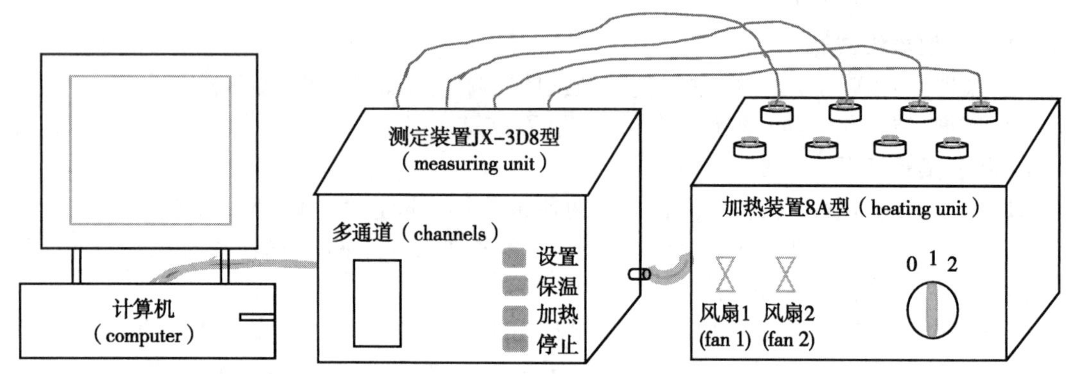

图 2-9-2　多通道金属相图测定装置

Fig. 2-9-2　Schematic of Multi-channel Metal Phase Diagram Measurement Apparatus

Experiment 9　Drawing of Phase Diagram for a Binary Eutectic System

1. Objective

1.1. To construct Cd-Bi phase diagram from cooling curves.

1.2. To master basic principles and application method of thermocouple thermometer.

1.3. To learn the pharmaceutical application of eutectic phase diagram.

2. Principles

Binary eutectic phase diagram is widely used in steel smelting, petrochemical, ceramic technology and product separation. A number of methods exist for constructing a phase diagram. The more prevalent one is thermal analysis method. Cooling the system from the high temperature under constant pressure, we can obtain the temperature-time curve, which is called step cooling curve. By measuring a series of cooling curves, we can construct the phase diagram of the tested system, as shown in Fig. 2-9-1.

Step cooling curves of pure substances, such as curve 1 and curve 5 in Fig. 2-9-1(a), shows quick decrease of temperature at first. The slope of line ab depends on the extent of heat dispersion of the cooling system. When the system cools down to the melting point of A, the liquid starts to solidify, the system exhibits two-phase equilibrium between solution and solid A, then the temperature remains unchanged until all the liquid has been converted to solid (horizontal segment bc). Then the temperature drops rapidly again.

Step cooling curves of mixtures (e. g. curve 2, curve 4) are different from those of pure substances(e.g. curve 1, curve 5). For example, curve 2 shows that the temperature quickly drops to point b′, at which solid starts to emerge and the system is two phases in equilibrium. Because solidifying changes the composition of the solution, the equilibrium temperate is continuously reduced. However, the heat of fusion is released during this process, which results in a slow dropping of the temperature and smaller slope of the curve (line b′c′). When the eutectic temperature is reached, both components solidify, the system exhibits three phases in equilibrium and the temperature holds constant again until all the liquid has been converted to solid (line c′d′). Then the temperature drops rapidly again.

Curve 3 represents exactly the shape of mixture equal to eutectic composition, which is similar to the graphics of the pure material, but its horizontal section is a three-phase balance.

Plotting temperature versus the composition of two components, the phase diagram can be constructed by connecting the transition temperature at specified composition.
Fig. 2-9-1(b) is a simple eutectic phase diagram of two-component system. Region L is the liquid area, left triangle represents two-phase region of pure solid A and liquid, right triangle represents two-phase region of pure solid B and liquid. The horizontal line shows coexistence of solid A, solid B and liquid phase. Below the horizontal line, both solid A and solid B coexist in equilibrium. O is the low eutectic point.

3. Apparatus and chemicals

3.1. Apparatus

500W heating furnace, thermocouples, ceramic crucibles, potentiometer, rigid tubes, transformer.

3.2. Chemicals

Cadmium(AR), bismuth(AR), paraffin oil.

4. Procedures

4.1. 50g of each sample with different composition is prepared and is respectively put in eight hard test tubes. Concentrations of Cd in weight percentage are respectively 0, 15%, 25%, 40%, 55%, 75%, 90% and 100%. Then add about 3g paraffin oil in each tube to prevent oxidation on air exposure of metal during heating.

4.2. Determine the step cooling curves of samples in order of pure Cd, pure Bi, 90% Cd, 75% Cd, 55% Cd, 40% Cd, 25% Cd and 15% Cd. Place the sample tube into the heating furnace (the voltage controlled at about 160V). Put the hot end of the thermocouple with glass sheath into the sample tube, while the cold end of the thermocouple in ice-water to heat the sample. When all the solids have been converted to liquids, stop heating. The melting points of Cd and Bi are used to calibrate the thermocouple. It is known that the melting points of Cd and Bi are 594.3K and 544.6K respectively.

4.3. Attentions

4.3.1. Control the cooling rate at 6~8K/min. The hot end of the thermocouple should be placed in the center of the sample and keep at least 1cm away from the bottom of the sample tube.

4.3.2. Operate carefully the sample tubes to prevent burns.

4.3.3. Mix the samples more uniformly.

4.3.4. A too fast cooling speed is not allowed to prevent an unobvious inflexion and plateau.

5. Data recording and processing

5.1. Construct a phase diagram of Cd-Bi binary system from cooling curves, and indicate the phase equilibrium in the designated regions.

5.2. Find the eutectic temperature and composition of the mixture.

6. Questions

6.1. What are the differences from the horizontal section in step cooling curves of different mixtures?

6.2. Whether can the heating curves be used to construct the phase diagram?

6.3. Are there any other methods for construction of phase diagram?

7. Pharmaceutical applications

Solid dispersions is an efficient means of improving absorption and bioavailability of drugs. When a mixture of insoluble drug and soluble carrier with eutectic composition is cooled rapidly, very fine crystals of the two components are released to form the solid dispersion system. The large surface area of the particles results in an enhanced dissolution rate, and consequently, improved bioavailability. For example, the dissolution of itraconazole from a eutectic mixture of the drug and poloxamer 188 was 4.66 times that of the pure drug and 2.96 times that of the physical mixture, and 12 times for a eutectic mixture of 48% urea and 52% sulfathiazole as compared with that of pure sulfathiazole.

8. Other methods

8.1. In order to facilitate the control of heating and cooling rate and observation of the changes of temperature for students, SWKY-1 type digital thermometer and KWL-09 type controllable heating and cooling furnace can be used to perform the experiment.

8.2. The original experimental sample tubes are self-made and can only measure one sample at a time. The device (e.g. JX-3D8) can be used to reduce the workload of the students. This device can measure 8 samples simultaneously, and the sample has been encapsulated and is reusable. Binary metal phase diagram can be obtained through inputting the inflection point and the platform to instrument software. The schematic diagram of multi-channel metal phase diagram measurement apparatus is shown in Fig. 2-9-2.

实验十 三组分液-液系统相图的绘制

（一）实验目的

1. 熟悉相律和用三角形坐标表示三组分相图的方法。
2. 绘制等温等压下具有一对共轭溶液的醋酸-氯仿-水三组分系统的相图。
3. 了解三元相图的药学应用。

（二）实验原理

根据相律，在恒温恒压下，可以用平面等边三角形图来表示三组分系统的平衡状态，如图 2-10-1 所示。等边三角形的三个顶点分别代表纯组分，三条边分别代表二组分系统，而三角形内任意一点,则代表三组分系统的组成。通常按逆时针方向在三角形三边上标出 A、B、C 三组分的质量分数(或摩尔分数)。以图 2-10-1 中 O 点为例,其组成可按下面方法求得:过 O 点作平行于三角形三边的直线 Oa、Ob、Oc,按几何学原理,Oa+Ob+Oc=AB=BC=CA,Oa、Ob、Oc 即分别为 A、B、C 的质量分数 w_A、w_B、w_C(或摩尔分数 x_A、x_B、x_C)。

图 2-10-2 所示的醋酸-水-氯仿三组分体系中,醋酸和水及醋酸和氯仿完全互溶,而水和氯仿只能有限度的互溶。图中 EOF 是溶解度曲线。溶解度曲线之内是共轭两相区,曲线上面是单相区。O_1、O_2 为物系点,K_1L_1、K_2L_2 称为连接线。当物系点从两相区转移到单相区,在通过相分界线 EOF 时,系统从浑浊变为澄清。从单相区转移到两相区通过相分界线 EOF 时,系统从澄清变为浑浊。因此,根据系统澄明度的变化,可以测定出溶解度曲线,绘出相图。

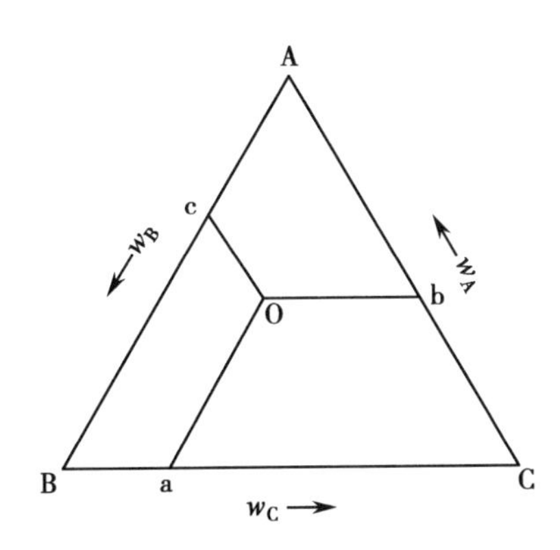

图 2-10-1 三角形坐标表示法
Fig. 2-10-1 Triangle Coordinate Denotation Method

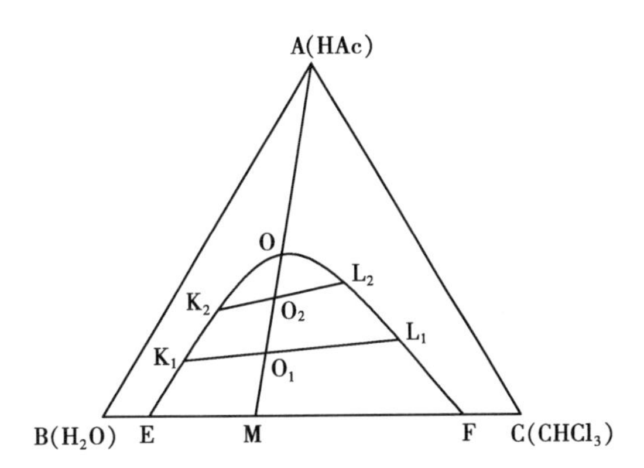

图 2-10-2 有一对共轭溶液的三组分系统相图
Fig. 2-10-2 Phase Diagram of a Three Component System

通常可用下述方法绘制溶解度曲线。配制以一定比例混合的 B 和 C 的部分互溶溶液,组成如图 2-10-2 中的 M 点。当不断加入液体 A 时,则系统组成沿 MA 线移动至 O 点,溶液由浑浊变为清澈,从而可以确定终点 O。继续加入少量 A,系统保持清澈,然后用氯仿滴定,系统出现浑浊时又会得到另一个终点。如此重复,则可画出溶解度曲线。

也可先配制完全互溶的 A 和 C 的混合溶液,然后向其中滴加水直至出现浑浊来确定终点,以画出溶解度曲线。本实验就采用此法。

(三)仪器和试剂

1. 仪器 25ml 和 100ml 磨口锥形瓶,100ml 锥形瓶,50ml 酸式滴定管,50ml 碱式滴定管,1ml 和 2ml 移液管,分液漏斗,漏斗架。

2. 试剂 氯仿(分析纯),冰醋酸(分析纯),0.5mol/L 标准 NaOH 溶液,酚酞指示剂。

(四)实验步骤

1. 如图 2-10-3 所示,在洁净的酸式滴定管内装入蒸馏水,在碱式滴定管内装入 NaOH 溶液。

2. 移取 10ml 氯仿及 4ml 醋酸于洁净的 100ml 磨口锥形瓶中,然后慢慢滴入水,同时不停振荡,至溶液恰好由清变浑,即为终点,记下所用水的体积。再向此瓶内加入 5ml 醋酸,系统又成均相,继续用水滴定至终点(由清变浑)。以后用同法加入 8ml 醋酸,用水滴定至终点，再加 8ml 醋酸，用水滴定，记录各次各组分的用量。最后再加入 40ml 水，加磨口塞，摇动，并每隔 5 分钟摇动一次，30 分钟后用此液体测连接线（溶液Ⅰ）。

另取一个洁净而干燥的 100ml 磨口锥形瓶，用移液管移入 1ml 氯仿及 3ml 醋酸，用水滴定至终点（由清变浑）。以后依次加 2ml、5ml、6ml 醋酸，分别用水滴定至终点，并记录各次各组分用量。最后再加入 10ml 氯仿和 5ml 醋酸，加磨口塞摇动，每隔 5 分钟摇动一次，30 分钟后用此液体测另一根连接线（溶液Ⅱ）。

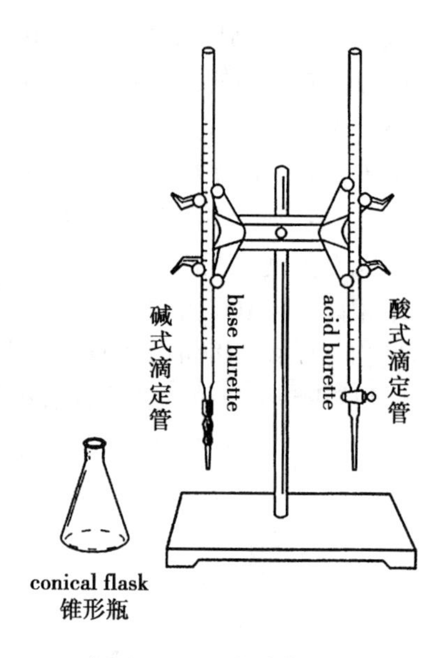

图 2-10-3　实验装置图

Fig. 2-10-3　Schematic Diagram of Experimental Equipment

3. 将溶液Ⅰ和溶液Ⅱ分别移至分液漏斗中，静置分层后，用洁净且干燥的移液管（或滴管）吸取溶液Ⅰ上层和下层液体各 2ml，分别置于已称量的 4 个 25ml 带塞磨口锥形瓶中，再称其质量，然后用水洗入 100ml 锥形瓶中，以酚酞为指示剂，用 0.5mol/L NaOH 标准溶液滴定以确定其中醋酸的含量。

同法吸取溶液Ⅱ上层和下层液体各 2ml，称重并滴定之，记录数据。

4. 清洗仪器，整理实验台。

5. 注意事项

（1）酸式滴定管易漏，所以试剂不宜久存管中。氯仿也可用刻度移液管加入。

（2）用水滴定如超过终点，则可再滴几滴醋酸至刚由浑浊变清作为终点，记下各组分的实际用量。在做最后几滴时（氯仿含量较少）终点是逐渐变化的，需滴至出现明显浑浊，才停止滴加水。

（3）在室温低于 16℃时，冰醋酸可恒温后用刻度移液管量取。

（4）用移液管（或滴管）吸取二相平衡的下层溶液时，可在吹气鼓泡条件下插入移液管，这样可避免上层溶液的玷污。

（五）数据记录和处理

1. 设计数据记录表格，记录终点时溶液中各组分的实际体积。由手册查出实验温度时三种液体的密度，算出各组分的重量百分含量，记入表中。

2. 在等边三角形坐标纸上，平滑地作出溶解度曲线。

（六）思考题

1. 如连接线不通过物系点，其原因可能是什么？

2. 为什么根据系统由清变浑的现象即可测定相界？

（七）药学应用

三元相图可以用来确定微乳的存在区域及微乳区面积大小，在药学研究中应用很广。例如在更昔洛韦自微乳化释药系统处方优化的研究中就利用了微乳的三元相图。更昔洛韦作为抗病毒药被广泛应用于预防和治疗免疫缺陷患者的巨细胞病毒感染。采用自微乳化技术，通过三元相图实验并结合乳液粒径的测定，可寻找出更昔洛韦最佳的处方配比。

Experiment 10 Drawing of Phase Diagram for a Ternary Liquid System

1. Objective

1.1. To master the Gibbs phase rule and the construction method of a phase diagram for a three-component system.

1.2. To draw the phase diagram for a three-component system containing acetic acid, chloroform and water on a triangular diagram at constant temperature and pressure.

1.3. To learn the pharmaceutical applications of ternary phase diagram.

2. Principle

According to Gibbs phase rule, it is convenient to represent a three-component system on a triangular diagram as illustrated in Fig.2-10-1 at constant T and p. In an equilateral triangle, any apex represents one component (pure compound) and the side, two components. The side is taken as 100% (percentage composition by mass) or as 1 (mole fraction). The composition of a mixture of three liquids can be represented by any point inside the triangle by measuring on these coordinates the distances toward apex of A, B and C respectively.

In the system illustrated in Fig. 2-10-2, pair A and B, and pair A and C are completely miscible at all concentrations, while pair B and C are partially soluble in each other. Curve EOF is called solubility curve, indicating the miscibility of B and C. All mixtures of the three liquids having compositions lying below the curve EOF will separate into two liquid phases, while all mixtures having compositions lying above the curve will give one homogeneous liquid phase. Points O_1 and O_2 are defined as system points, lines K_1L_1 and K_2L_2, the tie lines. Once the system point moves into the area of EOF from outside, the system will change from clearance to cloudiness when pass through the boundary of line EOF. Therefore, the solubility curve can be measured to draw the phase diagram by the clarity of the solution.

The solubility curve could be plotted in the following way. B and C are mixed at certain ratio to obtain a partially miscible liquid system. If component point starts at point M as shown in Fig. 2-10-2, it will move along line MA when sample A is kept adding into the system. The final point O is reached when system turns clear from cloudy. The solution is still clear if keeping adding sample A. Another final point is reached when performing titration with chloroform until solution becomes turbid again. The solubility curve is thus plotted by repeating these procedures.

The final point may also be obtained like this: prepare a homogeneous solution with sample A and C being mixed at certain ratio, then add A into the solution until the solution turns clear to cloudy. In this experiment, the solubility curve is obtained using this method.

3. Apparatus and chemicals

3.1. Apparatus

25ml and 100ml conical flasks with ground-in glass stopper, 100ml conical flask, 50ml acid burette, 50ml alkali burette, 1ml and 2ml pipettes, separatory funnel, funnel stand.

3.2. Chemicals

Chloroform(AR), glacial acetic acid(AR), 0.5mol/L standard sodium hydroxide(NaOH) solution, phenolphthalein indicator.

4. Procedures

4.1. Add water into acid burette, and NaOH solution into alkali burette.

4.2. Transfer 10ml of chloroform and 4ml of acetic acid using pipettes to a 100ml conical flask with ground-in glass stopper. Titrate the solution with water until the appearance of liquid from transparent to cloudiness and take it as end point. Record the volume of water consumed. Continuously add 5ml of acetic acid to this flask until a transparent solution appears. Again titrate with water to the end point. Repeat the titration after adding 8ml of acetic acid two more times. Record the volumes of water consumed in each titration. Then add 40ml of water to the flask, stopple the flask, and shake it every 5min for 30min. Take the solution as sample I for determination of tie line.

Transfer 1ml of chloroform and 3ml of acetic acid using pipettes to another 100ml conical flask with ground-in glass stopper, titrate the solution with water and observe the end point as described above. Record the volume of water consumed. Add 2ml, 5ml and 6ml of acetic acid in turn, repeat the titration as described above. Record the volume of water consumed in each titration. Then add 10ml of chloroform and 5ml of acetic acid, stopple the flask, and shake it every 5min for 30min. Take the solution as sample Ⅱ for determination of another tie line.

4.3. Transfer sample I into a separatory funnel. After a complete separation of two phases, respectively suck 2ml of the upper layer liquid and lower layer liquid using dry pipettes into four 25ml weighed conical flasks with ground-in glass stopper, weigh the flask again. Transfer the solution into 100ml conical flask completely, add phenolphthalein indicator, and titrate with 0.5mol/L standard NaOH solution to determine the concentration of acetic acid.

Repeat the procedure 4.3 to determine the concentration of acetic acid in sample Ⅱ.

4.4. Clean the instrument and tidy up the test bench.

4.5. Attentions

4.5.1. Long time storage of solution in acid burette is not allowed to prevent leakage of the solution.

4.5.2. If an over titration has happened in procedure 4.3, add a few drops of acetic acid until the appearance of liquid from cloudiness to transparent and take it as end point. Record the actual consumptions of each component. Titrate with water to an obvious cloudiness when the solution to be titrated contains small amount of chloroform.

4.5.3. Pipette the glacial acetic acid after being thermostated if the room temperature is lower than 16℃.

4.5.4. When pipette the equilibrium solution in lower layer, blow gently to exclude any drops of upper layer.

5. Data recording and processing

5.1. Draw a table to record the experiment data. Calculate the percentage by weight of chloroform, acetic acid and water, and record the data in the table.

5.2. Draw a phase diagram for chloroform, acetic acid and water on a triangular paper.

6. Questions

6.1. What is the reason if the tie lines do not pass through the point of system?

6.2. Why the change of appearance of liquid from transparent to cloudiness could be used to determine the boundary of the phase region?

7. Pharmaceutical applications

The ternary phase diagram contains information for the location and area of micro-emulsions, and thus has wide application in pharmaceutical research. For example, it is used in the optimization of formulation for the self-emulsifying drug delivery system of ganciclovir. Ganciclovir is an antiviral drug being widely used in the prevention and treatment of cytomegalovirus infection for immunocompromised patients. The self-emulsifying technique in combination with ternary phase diagram as well as the determination of emulsion particle size makes it possible to optimize the formulation of ganciclovir.

实验十一 电解质水溶液电导的测定及应用

（一）实验目的

1. 巩固溶液电导的基本概念，掌握测量电解质水溶液电导的原理与方法。
2. 测定弱电解质醋酸的电导率，并求摩尔电导率、解离度和解离常数。
3. 掌握电导滴定原理及操作，测定未知浓度的醋酸溶液。
4. 测定难溶盐的溶解度。
5. 了解电导测定的药学应用。

（二）实验原理

1. 电导基本概念　电解质溶液的导电能力用电导 G 表示，单位为 S（西门子）。电导与导体的截面积 A 成正比，与导体的长度 l 成反比，即

$$G=\kappa\frac{A}{l} \qquad \text{式(2-11-1)}$$

式中比例系数 κ 称为电导率，单位为 S/m。

式(2-11-1)可以写为

$$\kappa=G\frac{l}{A} \qquad \text{式(2-11-2)}$$

式中的 $\frac{l}{A}$ 对于一定的电导电极而言是一常数，称为电导电极常数，以 K 表示。则

$$\kappa=GK \qquad \text{式(2-11-3)}$$

在相距 1m 的两个平行电极之间，放置含有 1mol 电解质的溶液，此溶液的电导率称为摩尔电导率，用 Λ_m 表示，单位为 $S\cdot m^2/mol$。由于规定了电解质的量为 1mol，溶液的体积 V_m 将随浓度 c 而改变，即 $V_m=1/c$，c 的单位为 mol/m^3。所以，摩尔电导率 Λ_m 与电导率 κ 的关系为

$$\Lambda_m=\kappa V_m=\frac{\kappa}{c} \qquad \text{式(2-11-4)}$$

电解质的摩尔电导率随溶液浓度的稀释而增加，无限稀释时的摩尔电导率以 Λ_m^∞ 表示。对于强电解质溶液而言，其 Λ_m 和 c 的关系为

$$\Lambda_m = \Lambda_m^\infty - A\sqrt{c} \qquad \text{式(2-11-5)}$$

以 Λ_m 对 c 作图，外推至 $c=0$ 处可求得 Λ_m^∞。

2. 电导法测弱电解质的解离度和解离常数　对于弱电解质来说，某一浓度时的摩尔电导率与无限稀释时的摩尔电导率之比，表示该浓度下的解离度 α，即

$$\alpha = \frac{\Lambda_m}{\Lambda_m^\infty} \qquad \text{式(2-11-6)}$$

某一浓度时的摩尔电导率可通过实验测得，无限稀释时的摩尔电导率可查表得到。因此，可以用测定电导率的方法，计算该浓度下的解离度 α，进而可求得弱电解质的解离平衡常数。

如以醋酸为例，其解离平衡常数 K_c 与解离度 α 及浓度 c 之间的关系为

	HAc	$\rightleftharpoons$	H^+	+	Ac^-
起始时	c		0		0
平衡时	$c(1-\alpha)$		$c\alpha$		$c\alpha$

$$K_c = \frac{c^2\alpha^2}{c(1-\alpha)} = \frac{c\alpha^2}{1-\alpha} \qquad \text{式(2-11-7)}$$

3. 电导滴定　电导滴定是利用滴定终点前后溶液电导变化的转折来确定终点的方法。以 NaOH 滴定 HAc 溶液为例，在未滴定前，溶液是弱电解质 HAc，解离度很小，此时溶液的电导率很小。加入 NaOH 溶液后，OH^- 和 HAc 或 H^+ 结合成解离度极小的 H_2O，同时 Na^+ 和 HAc 中的 Ac^- 形成强电解质 NaAc，于是随着 NaOH 的加入溶液电导率逐渐上升。超过滴定终点后，继续加入 NaOH 溶液，过量的 OH^- 不再被 H^+ 结合成 H_2O，且因 OH^- 的导电能力也很强，大约是 Ac^- 的 5 倍。所以随着 NaOH 的加入溶液电导率迅速上升。因此，滴定终点前后溶液电导率的改变程度不同，滴定曲线为两条不同斜率的直线（图 2-11-1），它们的交点即为滴定终点。

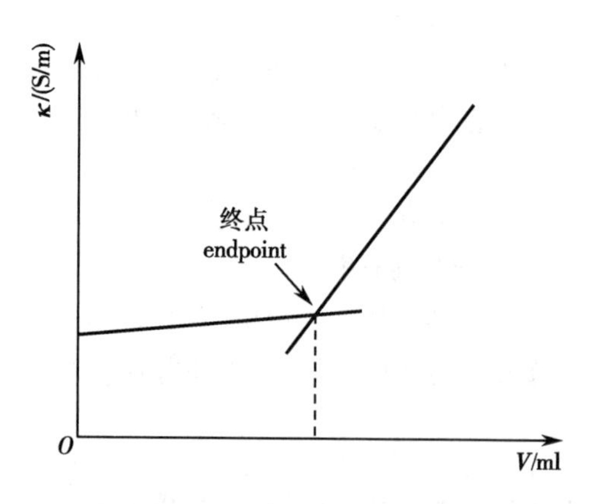

图 2-11-1　NaOH 滴定 HAc 的电导滴定曲线
Fig. 2-11-1　Conductimetric Titration Curve of HAc Titrated with NaOH

4. 电导法测难溶盐的溶解度　难溶盐氯化银的溶解度，也可通过测定其饱和水溶液的电导率而算出。

$$\kappa(\text{AgCl}) = \kappa(\text{饱和溶液}) - \kappa(H_2O) \qquad \text{式(2-11-8)}$$

由于难溶盐在水中的溶解度很小，溶液可视作无限稀，于是 AgCl 饱和水溶液的摩尔电导率可以用无限稀释摩尔电导率代替，即

$$\Lambda_m(\text{AgCl}) = \Lambda_m^\infty(\text{AgCl}) = \lambda_m^\infty(Ag^+) + \lambda_m^\infty(Cl^-) \qquad \text{式(2-11-9)}$$

根据式(2-11-4)，可计算得到氯化银在水中的溶解度 c

$$c=\frac{\kappa(\text{溶液})-\kappa(\text{水})}{\Lambda_m^\infty(\mathrm{AgCl})} \qquad 式(2\text{-}11\text{-}10)$$

(三)仪器和试剂

1. 仪器 DDS-11A 型电导率仪,电导电极,超级恒温水浴,50ml 烧杯,100ml 烧杯,移液管,碱式滴定管。

2. 试剂 0.020 0mol/L 标准 NaCl 溶液,准确配制 0.01mol/L 醋酸水溶液,0.1mol/L 标准 NaOH 溶液,AgCl 饱和水溶液。

(四)实验步骤

1. NaCl 无限稀释摩尔电导率测定 在烧杯中用移液管移入一定体积的 0.200mol/L 标准 NaCl 溶液,测定其电导率。然后用重蒸馏水依次稀释四次得到 0.010 0mol/L、0.005 00mol/L、0.002 50mol/L 和 0.001 25mol/L 的 NaCl 溶液,分别测其电导率值。

2. 醋酸解离常数的测定 将 50ml 烧杯与电导电极依次用蒸馏水及待测的醋酸水溶液冲洗两次,然后装入被测的醋酸溶液,插入电导电极。在 25℃ 恒温水浴中,恒温 10 分钟后测定其电导率,重复测定三次。

3. 醋酸溶液的浓度测定 取干净的 100ml 烧杯,用移液管加入 25ml 未知浓度的醋酸溶液,插入电导电极测其电导率。然后用滴定管每次加入 0.1mol/L NaOH 标准溶液 2~3ml,待搅拌均匀后,测其电导率,过滴定终点后再做 6~7 个实验点即告结束,列表记录数据。

4. 难溶盐氯化银在水中的溶解度测定 将 50ml 烧杯与电导电极依次用蒸馏水及待测的氯化银饱和溶液冲洗两次,然后装入被测的氯化银饱和溶液,插入电导电极。在 25℃ 恒温水浴中,恒温 10 分钟后测定其电导率,重复测定三次。同法测定蒸馏水的电导率。

(五)数据记录与处理

1. 已知:$\Lambda_m^\infty(\mathrm{HAc})=0.039\ 07\mathrm{S\cdot m^2/mol}$,$\Lambda_m^\infty(\mathrm{AgCl})=0.013\ 826\mathrm{S\cdot m^2/mol}$。

2. 根据实验步骤 1 设计数据记录表格,以 Λ_m 对 c 作图,求 NaCl 的 Λ_m^∞。

3. 根据实验步骤 2 设计数据记录表格,将实验数据列入表格中,求 HAc 的解离常数。

4. 根据实验步骤 3 的数据记录,将 HAc 溶液的电导率对加入 NaOH 溶液的体积作图求得滴定终点,并计算 HAc 溶液的浓度。

5. 根据实验步骤 4 的数据记录,由式(2-11-10)求出 AgCl 在水中的溶解度 c。

(六)思考题

1. 影响弱电解质溶液电导率的因素有哪些?

2. 电导率测定中对使用的水有什么要求?

3. 测定溶液的电导率有何实际应用?

4. 为什么滴定液浓度比待测液大至少十倍?若浓度太小有何影响?

(七)药学应用

电导率测定在药学实践中有着重要的意义。例如,纯净水在药品生产中被用作辅料、溶剂或分析试剂等,因此电导率测定在药品生产企业或药品研究单位被用来对水质进行控制和分析。电导率测定也可用于鉴别乳状液类型、乳状液转型和稳定性的研究。

(八)电导率测量技术及仪器使用说明

1. DDS-11A 型电导率仪的操作要点 开机预热,检查指针是否指零。确定“常数”,选择工作频率(电导率>10^3 挡,用“高周”,否则用“低周”),选择适当量程挡。打开“校正”开关,旋动“调整”旋钮,将指针调至满刻度,最后“测量”开关打开,读数。量程选择应由大到小，注意红挡和黑挡量程在表盘中读数的区别。

2. DDS-11A 型电导率仪的使用注意事项

（1）电极引线不能潮湿，否则所测数据不准。

（2）纯水的测量要迅速，因为空气中的二氧化碳溶解在水中，导致水的电导率迅速增加，影响测量结果。

（3）盛待测液的容器必须清洁，无离子污染。

（4）擦拭电极时不可触及铂黑，以免铂黑脱落，引起电极常数的改变。

Experiment 11 Measurement and Application of Electrolyte Solution Conductance

1. Objective

1.1. To get familiar with the concept of conductance, to learn the principle and the method of conductance measurement for aqueous electrolyte solution.

1.2. To measure the molar conductance and to calculate the dissociation degree and dissociation constant of weaker electrolyte by conductimetric method.

1.3. To learn the principle and the operation of conductimetric titration, to determine the concentration of acetic acid solution.

1.4. To determine the solubility of slightly soluble salt.

1.5. To learn the pharmaceutical application of conductivity measurement.

2. Principle

2.1. Concepts of conductance

Conductance(G) is a measure of the ability of electrolytic solutions to conduct electricity, and the unit of conductance G is S(Siemens). It is the reciprocal of resistance and is proportional to the cross-sectional area A and the thickness l of the conductor, that is

$$G=\kappa\frac{A}{l} \tag{2-11-1}$$

where κ is the specific conductance or conductivity(S/m).

Therefore, equation(2-11-1) can be expressed as

$$\kappa=G\frac{l}{A} \tag{2-11-2}$$

As will be seen later, the ratio l/A is usually treated as an apparatus or cell constant and is given the symbol of K. Then, we have

$$\kappa=GK \tag{2-11-3}$$

The molar conductivity, Λ_m($S \cdot m^2/mol$) is the conductivity of an electrolyte solution at any particular concentration by imagining two large parallel electrodes set 1m apart and supposing the whole of the solution containing 1 mole of electrolyte to be placed between them. So we have

$$\Lambda_m=\kappa V_m=\frac{\kappa}{c} \tag{2-11-4}$$

where c is the concentration of the solution in mole per cubic meter.

The molar conductivity of electrolyte increases with increasing dilution. The molar conductivity at zero concentration, Λ_m^∞, is called molar conductivity at infinite dilution or limiting molar conductivity. For a strong electrolyte solution, the relationship between Λ_m and c is as follows

$$\Lambda_m = \Lambda_m^\infty - A\sqrt{c} \qquad (2\text{-}11\text{-}5)$$

Plotting Λ_m versus c, the Λ_m^∞ can be obtained from the intercept at $c=0$ of the curve.

2.2. Determinations of dissociation degree and dissociation constant of weak electrolyte

For a weak electrolyte, the ratio of molar conductivity at a specified concentration to the limiting molar conductivity represents the degree of dissociation α. That is

$$\alpha = \frac{\Lambda_m}{\Lambda_m^\infty} \qquad (2\text{-}11\text{-}6)$$

The molar conductivity at a specified concentration can be determined by the experiments and the molar conductivity at infinite dilution can be obtained from physical chemistry manual. By determination of the conductivity of electrolyte, the degree of dissociation α can be calculated and then the dissociation constant K_c can be calculated as well.

Considering the dissociation equilibrium of acetic acid, the relationship between K_c and α is

$$\begin{array}{cccc} HAc & \rightleftharpoons H^+ & + & Ac^- \\ c(1-\alpha) & c\alpha & & c\alpha \end{array}$$

$$K_c = \frac{c^2\alpha^2}{c(1-\alpha)} = \frac{c\alpha^2}{1-\alpha} \qquad (2\text{-}11\text{-}7)$$

2.3. Conductimetric titration

Conductimetric titration is the use of conductance change before and after titration endpoint to determine the end point. The principle of conductimetric titration can be illustrated by the titration of acetic acid (HAc) solution with sodium hydroxide (NaOH) solution. Since the HAc is a weak acid, the conductivity is initially small. As NaOH is added, OH^- from the base reacts with the HAc or H^+ from the acid to form neutral water molecules, Na^+ reacts with Ac^- to form the strong electrolyte NaAc. The conductivity of the solution increases gradually. After the endpoint, addition of NaOH solution again results in an increase in concentration of OH^- in solution and the relative conductivity of OH^- is five times that of the Ac^-, thus the conductivity increases more rapidly than it did before the end point. So the titration curve is composed of two straight lines of different slopes due to the different change of solution's conductivity before and after titration endpoint. Then the intersection point of these two straight lines is the titration end point (Fig.2-11-1).

2.4. Determination of solubility of slightly soluble salt

The solubility of slightly soluble salt, silver chloride (AgCl), in water, can be measured by conductimetric method. The conductivities of pure water and a saturated solution of AgCl in pure water are measured. The net value of conductivity of the slightly soluble salt is

$$\kappa(AgCl) = \kappa(\text{saturated}) - \kappa(H_2O) \qquad (2\text{-}11\text{-}8)$$

As the solubility of the slightly soluble salt is small, the solution is considered infinite dilution. And the molar conductivity of the salt can be assumed to be equal to the infinite dilution value, $\Lambda_m = \Lambda_m^\infty$, then we have

$$\Lambda_m(AgCl) = \Lambda_m^\infty(AgCl) = \lambda_m^\infty(Ag^+) + \lambda_m^\infty(Cl^-) \qquad (2\text{-}11\text{-}9)$$

And hence the solubility of AgCl may be calculated from the following equation

$$c=\frac{\kappa(\text{solution})-\kappa(\text{water})}{\Lambda_m^\infty(\text{AgCl})} \tag{2-11-10}$$

3. Apparatus and Chemicals

3.1. Apparatus

DDS-11A conductometer, conductance electrode, super thermostatic water bath, 50ml beaker, 100ml beaker, pipette, alkali burette.

3.2. Chemicals

0.020 0mol/L standard NaCl solution, 0.01mol/L HAc solution, 0.1mol/L standard NaOH solution, saturated solution of AgCl.

4. Procedures

4.1. Determination of Λ_m^∞ of NaCl

Pipette a specified volume of 0.200mol/L standard NaCl solution, measure the conductivity. Dilute the NaOH solution with distilled water to the concentrations of 0.010 0mol/L, 0.005 00mol/L, 0.002 50mol/L and 0.001 25mol/L, and measure their conductivities, respectively.

4.2. Determination of dissociation constant of HAc solution

Rinse the 50ml beaker and electrode two times each with water and 0.010 0mol/L HAc. Add HAc solution in the beaker, and immerse the electrode into the solution. After a stable temperature at 25℃ is reached, measure the conductivity. Repeat the measurement for three times.

4.3. Determination of the concentration of HAc solution

Add 25ml of HAc solution to a 100ml beaker using the pipette and measure the initial conductivity of the solution under stirring. Fill the alkali burette with 0.1mol/L standard NaOH solution. Add about 2~3ml of the standard NaOH solution to the beaker. Record the conductivity of the titrand and the total volume of added NaOH. Continue the titration until 6~7 data have been obtained after the titration end point.

4.4. Determination of solubility of AgCl in water

Rinse the 50ml beaker and electrode two times each with water and saturated solution of AgCl, respectively. Add saturated solution of AgCl into the beaker, and immerse the electrode into the solution. After a stable temperature at 25℃ is reached, measure the conductivity. Repeat the measurement for three times. Substitute water for the saturated salt solution and repeat the procedure described above.

5. Data recording and processing

5.1. It is known that $\Lambda_m^\infty(\text{HAc})=0.039\ 07\text{S}\cdot\text{m}^2/\text{mol}$, $\Lambda_m^\infty(\text{AgCl})=0.013\ 826\text{S}\cdot\text{m}^2/\text{mol}$.

5.2. Plot Λ_m versus c to obtain $\Lambda_m^\infty(\text{NaCl})$.

5.3. Draw a table to record the experimental data and calculate the dissociation constant of HAc solution.

5.4. Draw the titration curve to obtain the titration end point and calculate the concentration of HAc solution.

5.5. Use equation (2-11-10) to calculate the solubility of AgCl in water.

6. Questions

6.1. What factors will influence the conductivity of weak electrolyte solution?

6.2. What is the requirement to water in conductivity measurement?

6.3. To what fields can conductimetric method be applied?

6.4. Why the concentration of titrant should be 10 times higher than that of the titrand? What would be the effect on the result if a dilute titrant solution is used?

7. Pharmaceutical Applications

Conductivity measurement plays an important role in pharmaceutical practice. Purified water is used as an excipient, solvent in the processing, formulation and manufacture of pharmaceutical products, analytical reagents and in other pharmaceutical application. The use of conductivity as a quality control and diagnostic tool by pharmaceutical facilities and even small laboratory are common. Conductivity measurement is also a technique to classify the type of emulsion, study the stability of the emulsion or track emulsion inversion.

8. Instrument usage

8.1. Operation of DDS-11A type conductometer

Preheat the instrument for more than 10min, and check whether there is a zero display. Choose the cell constant according to the conductance electrode used. Select the working frequency. Choose "high frequency", if the conductivity is higher than 10^3, otherwise choose the "low frequency". Set the switch to position of "calibration", turn "adjust" to have a full scale. Then set the switch to "measurement", read the display and record it. The selection of measurement range should be from large to small. Pay attention to the display difference between the measurement range in red and in black.

8.2. Keys to operation of DDS-11A conductometer

(1) Do not wet the electrode wire otherwise you will get inaccurate data.

(2) A quick measurement of the conductivity for water is important, because of the dissolution of CO_2 in water which may lead to a quick increase of the conductivity of water and influence the experimental results.

(3) All the containers should be clean and with no ion contaminated.

(4) Take care not to touch the platinized surface of the electrode when wipe the electrode, or the platinum black will be broken off and result in a variation in cell constant.

实验十二 电动势法测溶液 pH 和反应热力学函数

（一）实验目的

1. 学会一些电极的制备和处理方法。
2. 测定电池的电动势和溶液的 pH。
3. 掌握电动势法测定化学反应热力学函数的原理和方法。
4. 掌握对消法的原理及电位差计的测量原理和使用方法。
5. 了解 pH 测定的药学应用。

（二）实验原理

1. 电动势测定原理 为了使电池反应在接近热力学可逆条件下进行，电池电动势必须用电位差计来测量。电位差计是采用对消法（或称补偿法）原理，在无电流通过的情况下测定电动势的，其工作原理见图 2-12-1 所示。

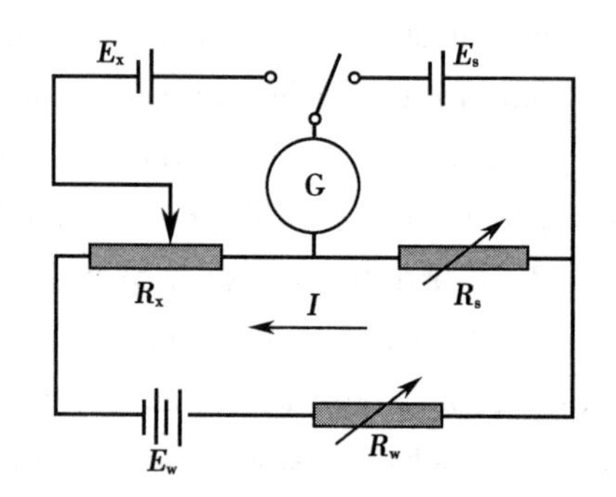

图 2-12-1 电位差计工作原理示意图
Fig. 2-12-1 A Potentiometer Circuit

首先根据实验温度，调整标准电阻 R_s 值，接通标准电池 E_s，调整工作电阻 R_w，使检流计 G 指零，即对消 E_s，此时工作电流是个定值：$I=\frac{E_s}{R_s}=$常数。然后接通待测电池 E_x，调整电阻 R_x，使检流计 G 指零，此时待测电池的电动势 E_x 等于电阻 R_x 两端的电压降 $E_x=IR_x=$ 常数$\times R_x$，R_x 在电位差计上直接表示为电动势值。

2. pH 的测定 利用各种氢离子指示电极与参比电极组成电池，通过电动势的测定可计算得到溶液的 pH。本实验以醌-氢醌（Q-H_2Q）电极作指示电极，饱和甘汞电极作参比电极，组成电池，即

饱和甘汞电极 ‖ 未知 pH 溶液 | Q-H_2Q | Pt(s)

醌-氢醌的电极反应为

$$C_6H_4O_2+2H^++2e^-\longrightarrow C_6H_4(OH)_2$$

其电极电势为

$$\varphi_{Q/H_2Q}=\varphi^{\ominus}_{Q/H_2Q}-\frac{RT}{2F}\ln\frac{a_{H_2Q}}{a_Q a^2_{H^+}} \qquad 式(2\text{-}12\text{-}1)$$

因为溶液中 Q-H_2Q 的浓度相等，且稀溶液中活度系数大致相等，因此 $a_Q=a_{H_2Q}$。则醌-氢醌的电极电势表示为

$$\varphi_{Q/H_2Q}=\varphi^{\ominus}_{Q/H_2Q}-\frac{RT}{2F}\ln\frac{1}{a^2_{H^+}}=\varphi^{\ominus}_{Q/H_2Q}-\frac{2.303RT}{F}\text{pH} \qquad 式(2\text{-}12\text{-}2)$$

这样，上述电池的电动势为

$$E=\varphi^{\ominus}_{Q/H_2Q}-\frac{2.303RT}{F}\text{pH}-\varphi_{甘汞(饱和)} \qquad 式(2\text{-}12\text{-}3)$$

因此，若测得电池的电动势，溶液的 pH 可按下式求得，即

$$\text{pH}=\frac{\varphi^{\ominus}_{Q/H_2Q}-\varphi_{甘汞(饱和)}-E}{2.303RT/F} \qquad 式(2\text{-}12\text{-}4)$$

在实际测量中，为了消除或减少液体接界电势及所制甘汞电极的电极电位的不准，通常采用二次测量法。即首先测量已知 pH 的标准缓冲溶液作为电极液时的电池电动势 E_s，然后测量待测溶液作为电极液时的电池电动势 E_x，此时溶液 pH 计算公式为

$$\text{pH}_x=\text{pH}_s+\frac{E_s-E_x}{1.984\times10^{-4}T} \qquad 式(2\text{-}12\text{-}5)$$

3. 测定可逆电池的电动势求算电池反应的热力学函数 在可逆电池中，恒温恒压下，化学反应的摩尔吉布斯能变化 Δ_rG_m 与原电池电动势的关系为

$$\Delta_rG_m=-zFE \qquad 式(2\text{-}12\text{-}6)$$

式中 z 为电池反应中的电子计量系数；F 为法拉第常数；E 为反应温度下可逆电池的电动势。

一定温度下，电池反应的摩尔熵变化 Δ_rS_m，可根据吉布斯-亥姆霍兹公式求得，即

$$\left[\frac{\partial(\Delta_r G_m)}{\partial T}\right]_p = -\Delta_r S_m \qquad 式(2\text{-}12\text{-}7)$$

将式(2-10-6)代入上式，得

$$\Delta_r S_m = zF\left(\frac{\partial E}{\partial T}\right)_p \qquad 式(2\text{-}12\text{-}8)$$

式中$\left(\frac{\partial E}{\partial T}\right)_p$称为电池电动势的温度系数，可通过测定电池在不同温度下的电动势求得。

反应的摩尔焓变 $\Delta_r H_m$，可由热力学关系式 $\Delta_r G_m = \Delta_r H_m - T\Delta_r S_m$ 和式(2-12-6)及式(2-12-8)求出，即为

$$\Delta_r H_m = -zEF + zFT\left(\frac{\partial E}{\partial T}\right)_p \qquad 式(2\text{-}12\text{-}9)$$

本实验测定的化学反应为

$$Ag(s) + \frac{1}{2}Hg_2Cl_2(s) \longrightarrow AgCl(s) + Hg(l)$$

将该反应设计成电池，即为

$$Ag(s)|AgCl(s)|KCl(0.1mol/L)|Hg_2Cl_2(s)|Hg(l)$$

虽然该电池的电动势与 KCl 溶液的活度无关，即 $E = E^{\ominus}$，但是，KCl 浓度太大对银-氯化银电极有溶解作用，故本实验采用 0.1mol/L 的 KCl 作电极液。

（三）仪器和试剂

1. 仪器　UJ-25 型电位差计，检流计，标准电池，直流稳压电源（或干电池），半电池管，盐桥，饱和甘汞电极，甘汞电极（0.1mol/L KCl），Zn 电极，Cu 电极，铂电极，银-氯化银电极，超级恒温水浴，恒温隔套，磁力搅拌器，500ml 和 50ml 烧杯。

2. 试剂　0.1mol/L $ZnSO_4$，0.1mol/L $CuSO_4$，醌-氢醌溶液，待测溶液，稀硝酸，稀硫酸（约 3.0mol/L），饱和 $Hg_2(NO_3)_2$ 溶液。

（四）实验步骤

1. 电极制备

（1）锌电极的制备：对于 Zn 电极需要进行汞齐化，以稀硫酸浸洗 Zn 棒大约 30 秒后，用蒸馏水冲洗，再将其浸入 $Hg_2(NO_3)_2$ 溶液中 3~5 秒，取出后用滤纸轻轻擦亮其表面，然后用蒸馏水洗净。汞齐化时注意汞有毒，所用过的滤纸应丢在废纸缸中，绝不允许随便丢在地上。取一个洁净的半电池管，插入已处理的电极金属，并塞紧封口使不漏气，然后由支管吸入 0.1mol/L $ZnSO_4$ 溶液即可。

（2）铜电极的制备：用稀硝酸洗净铜棒表面的氧化物，用蒸馏水冲洗。以该铜电极作阴极，另取一铜棒作阳极，在镀铜溶液中电镀约 15 分钟（电流密度约为 $20mA/cm^2$）。电镀完成后，分别用蒸馏水和 0.1mol/L $CuSO_4$ 冲洗，然后浸入 0.1mol/L $CuSO_4$ 溶液中即得铜电极。

（3）醌-氢醌电极的制备：取少量醌氢醌固体加入标准缓冲溶液或待测溶液中，搅拌使其成为饱和溶液，插入清洗干净的铂电极即为醌-氢醌电极。

2. 电动势的测定

（1）根据表 2-12-1 组成电池，将电位差计处于“断”位，按图 2-12-2 连接各装置并按下述方法测定各电池的电动势。

（2）记录实验温度，按下式计算实验温度时的 E_s，并在电位差计的适当位置调整到计

算值。

$$E_s = 1.01845 - 4.05 \times 10^{-5}(T-293) - 9.5 \times 10^{-7}(T-293)^2 + 1 \times 10^{-8}(T-293)^3 \quad 式(2\text{-}12\text{-}10)$$

（3）接通标准电池，调整工作电阻钮 R_W（由粗到微），按下检流计按钮（先粗后细）查看指针的偏转，直到检流计指零。

（4）接通待测电池（X_1 或 X_2 挡），调整电阻 R_x 钮（由大到小），按下检流计按钮（先粗后细）查看指针的偏转，直到检流计指零。然后读取待测电池的电动势值。

3. 反应的热力学函数测定

（1）如图 2-12-3 所示连接各装置和接线，开启磁力搅拌器。

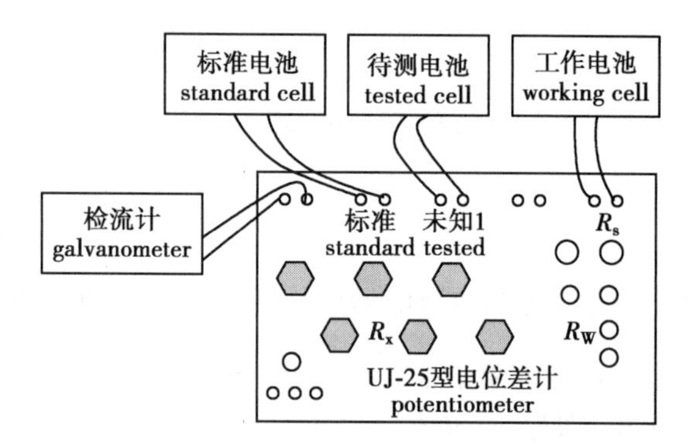

图 2-12-2 电动势测定装置图

Fig. 2-12-2 Sketch of the Apparatus for Electromotive Force Measuring

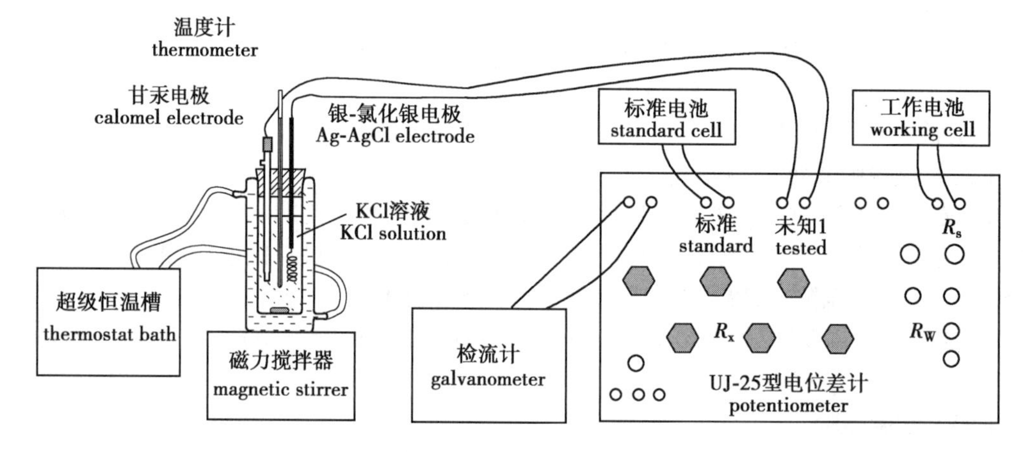

图 2-12-3 热力学函数测定装置示意图

Fig. 2-12-3 Sketch of the Apparatus for Thermodynamic Function Measuring

（2）开启超级恒温槽，调节温度为定值（如 298K），在电池充分恒温后进行电动势测定。

（3）调节超级恒温槽，使温度上升 5℃左右，待充分恒温后，测定该温度下的电动势。再次调高超级恒温槽温度约 5℃进行测定，共测定 5~6 个数据，温度读数精确到 0.1℃，估读到 0.02℃。

（五）数据记录与处理

1. 按表 2-12-1 和表 2-12-2 记录数据。

表 2-12-1 电池组成及电动势测定数据

Table 2-12-1 Cell Composition and Data of Electromotive Force

电池（cell）	E/V
（1）Zn | $ZnSO_4$（0.1mol/L）‖ 饱和甘汞电极（saturated calomel electrode）	
（2）饱和甘汞电极（saturated calomel electrode）‖ $CuSO_4$（0.1mol/L）| Cu	

续表

电池（cell）	E/V
（3）Zn | $ZnSO_4$（0.1mol/L）‖ $CuSO_4$（0.1mol/L）| Cu	
（4）饱和甘汞电极 ‖ 标准缓冲溶液 | Q-H_2Q |Cu （saturated calomel electrode ‖ standard buffer solution | Q-H_2Q |Cu）	
（5）饱和甘汞电极 ‖ 待测溶液 | Q-H_2Q |Cu （saturated calomel electrode ‖ solution to be tested | Q-H_2Q |Cu）	

表 2-12-2 不同温度下的电池电动势测定值

Table 2-12-2 Electromotive Forces at Different Temperatures

标准电池温度（temperature of standard cell）：________℃；

标准电池电动势（electromotive force of standard cell）：________ V

No.	t/℃	T/K	E/V
1			
…			
5			

2. 根据式（2-12-5）计算待测液 pH。

3. 作 $E \sim T$ 图，并对 $E \sim T$ 线性回归，计算电池电动势温度系数 $\left(\frac{\partial E}{\partial T}\right)_p$。

4. 根据 $E \sim T$ 关系式，计算 298K 时电池的电动势 E，电池反应的 $\Delta_r G_m$、$\Delta_r S_m$ 和 $\Delta_r H_m$。

（六）思考题

1. 如果待测电极极性接反会出现什么现象？线路未接通又会出现什么现象？
2. 锌电极为何要制作成锌汞齐电极，这样处理对锌电极电势是否有影响？
3. 为什么要采用对消法测定电池电动势？
4. 如何用电动势法测定电池反应的平衡常数？
5. 测定电池电动势的实验装置中，工作电池、标准电池和检流计各起什么作用？

（七）药学应用

在药学研究中，系统的 pH 控制非常重要。如很多药物的降解属氧化还原反应，有氢离子参与。因此，药物降解与氢离子有关。例如，阿扑吗啡在酸性溶液中可以稳定存在，但是在碱性或中性溶液中很快降解。通常，通过控制 pH 可以防止液体制剂的氧化降解。

Experiment 12 Determination of pH of Solutions and Thermodynamic Functions by Electromotive Force Measurements

1. Objective

1.1. To learn the preparation of several electrodes.

1.2. To determine pH of the solution by electromotive force（EMF）measurements.

1.3. To master the principle and methods for determination of thermodynamic functions for chemical reactions by EMF measurements.

1.4. To master the principle of compensation technique for EMF measurement and the usage of a potentiometer.

1.5. To learn the pharmaceutical application of pH determination.

2. Principle

2.1. EMF measurements

The EMF of a cell can be determined with a potentiometer which employs the compensation technique devised by Poggendorff. The fundamental principle of a potentiometer is shown in Fig. 2-12-1.

Under a specified experimental temperature, adjust the standard resistor R_s, set the slide wire to the standard cell, E_s, then adjust the working resistor R_w until no current flow through the galvanometer G. The working current in this case is constant and is given by $I=\frac{E_s}{R_s}=\text{constant}$. Next, set the slide wire to the cell to be measured, E_x, adjust the resistor of R_x until no current flow through the galvanometer G. The potential value is now given by $E_x=IR_x=\text{constant}\times R_x$. Then we can read the potential value directly from the potentiometer.

2.2. Determination of pH of a solution

The pH of a solution can be calculated by determination of the EMF of a cell which is composed of a pH sensitive electrode and a reference electrode. In this experiment, a quinhydrone ($Q\text{-}H_2Q$) electrode is selected as the pH sensitive electrode, a saturated calomel electrode (SCE) as the reference electrode. The cell is

$$Hg(l)\,|\,Hg_2Cl_2(s)\,|\,KCl(\text{saturated})\,\|\,\text{tested solution}\,|\,Q\text{-}H_2Q(aq)\,|\,Pt(s)$$

The electrode reaction for $Q\text{-}H_2Q$ electrode is

$$C_6H_4O_2+2H^++2e^-\longrightarrow C_6H_4(OH)_2$$

The potential of $Q\text{-}H_2Q$ electrode can be calculated using Nernst equation, that is

$$\varphi_{Q/H_2Q}=\varphi^{\ominus}_{Q/H_2Q}-\frac{RT}{2F}\ln\frac{a_{H_2Q}}{a_Q a^2_{H^+}} \tag{2-12-1}$$

In aqueous solution, the concentrations of quinone (Q) and hydroquinone (H_2Q) are the same and both have the same activity coefficients. Thus, the electrode potential becomes

$$\varphi_{Q/H_2Q}=\varphi^{\ominus}_{Q/H_2Q}-\frac{RT}{2F}\ln\frac{1}{a^2_{H^+}}=\varphi^{\ominus}_{Q/H_2Q}-\frac{2.303RT}{F}\text{pH} \tag{2-12-2}$$

Now, the EMF of the cell is given by

$$E=\varphi^{\ominus}_{Q/H_2Q}-\frac{2.303RT}{F}\text{pH}-\varphi_{SCE} \tag{2-12-3}$$

If the EMF of the cell is obtained, the pH of the tested solution can be calculated by

$$\text{pH}=\frac{\varphi^{\ominus}_{Q/H_2Q}-\varphi_{SCE}-E}{2.303RT/F} \tag{2-12-4}$$

Usually a standard buffer with known pH is used for measuring of E_s, then the tested solution is used for measuring of E_x. Therefore, the EMF of the tested solution can be calculated using the following equation

$$pH_x = pH_s + \frac{E_s - E_x}{1.984 \times 10^{-4} T} \qquad (2\text{-}12\text{-}5)$$

2.3. Thermodynamics of reversible chemical cell

The EMF of a reversible chemical cell depends on the temperature. From the temperature coefficient of EMF and knowledge of EMF at a particular temperature, the Gibbs energy change, $\Delta_r G_m$, enthalpy change, $\Delta_r H_m$, and entropy change, $\Delta_r S_m$ for the cell reaction can be calculated with the aid of the following equations

$$\Delta_r G_m = -zFE \qquad (2\text{-}12\text{-}6)$$

$$\left[\frac{\partial(\Delta_r G_m)}{\partial T}\right]_p = -\Delta_r S_m \qquad (2\text{-}12\text{-}7)$$

$$\Delta_r S_m = zF\left(\frac{\partial E}{\partial T}\right)_p \qquad (2\text{-}12\text{-}8)$$

$$\Delta_r H_m = -zEF + zFT\left(\frac{\partial E}{\partial T}\right)_p \qquad (2\text{-}12\text{-}9)$$

where z is the number of electrons transferred in the reaction, F is the Faraday constant, E is the EMF of the cell, and $(\partial E/\partial T)_p$ is the temperature coefficient of EMF.

The chemical reaction studied in this experiment is as follows

$$Ag(s) + \frac{1}{2}Hg_2Cl_2(s) \longrightarrow AgCl(s) + Hg(l)$$

The schematic cell diagram is

$$Ag(s) | AgCl(s) | KCl(0.1mol/L) | Hg_2Cl_2(s) | Hg(l)$$

Although the electromotive force of the cell has nothing to do with the activity of KCl solution, that is, $E = E^{\ominus}$, the concentration of KCl is too large to have a dissolving effect on the silver-silver chloride electrode, so 0.1mol/L KCl was used as the electrode solution in this experiment.

3. Apparatus and chemicals

3.1. Apparatus

UJ-25 potentiometer, galvanometer, standard cell, DC power supply (or batteries), half battery tube, salt bridge, saturated calomel electrode, calomel electrode (0.1mol/L KCl), Zn electrode, Cu electrode, platinum electrode, silver-silver chloride electrode, super thermostat bath, thermostatic sleeve, magnetic stirrer, 500ml and 50ml beakers.

3.2. Chemicals

0.1mol/L $ZnSO_4$, 0.1mol/L $CuSO_4$, Q-H_2Q solution, tested solution, dilute nitric acid, dilute sulfuric acid, saturated mercury nitrate solution.

4. Procedures

4.1. Preparation of electrodes

4.1.1. Preparation of Zinc electrode: Immerse the zinc rod into a dilute sulfuric acid solution to remove the oxide film and dirt for 30s. Wash it with distilled water, and then put it into saturated solution of mercury nitrate for 3~5s. Wash it with distilled water again. Place the treated zinc rod into a clean half-cell tube, and then inhale 0.1mol/L ZnSO solution from the side tube to obtain the Zn electrode.

4.1.2. Preparation of copper electrode: Immerse the copper rod into a dilute nitric acid solution to remove the oxide film and dirt, and wash it with distilled water. Then electroplate the copper rod for about 15min. Wash the treated copper rod with distilled water and 0.1mol/L $CuSO_4$ solution. Then insert it into 0.1mol/L $CuSO_4$ solution to obtain the Cu electrode.

4.1.3. Preparation of Q-H_2Q electrode: Place a small amount of Q-H_2Q solid into a saturated buffer solution or tested solution. Stir it to form a saturated solution. Insert a clean platinum electrode into the saturated Q-H_2Q solution to obtain the Q-H_2Q electrode.

4.2. Measurement of EMF

4.2.1. Switch the potentiometer to the "off" position. Compose the cells accord to table 2-10-1 and connect all parts of the apparatus as shown in Fig. 2-12-2.

4.2.2. Record the experimental temperature, calculate EMF of the standard cell at this temperature by using equation (2-12-10), and adjust R_S to the corresponding value.

$$E_s = 1.018\,45 - 4.05\times10^{-5}(T-293) - 9.5\times10^{-7}(T-293)^2 + 1\times10^{-8}(T-293)^3 \qquad (2\text{-}12\text{-}10)$$

4.2.3. Set switch to standard cell, adjust the working resistor R_w from rough to fine, and press the galvanometer button from rough to fine until no current cross the galvanometer.

4.2.4. Set switch to the cell to be measured, adjust the resistor R_x from big button to small button, then press the buttons for galvanometer adjustment from rough to fine until no current cross the galvanometer. Read out the EMF displayed in the windows of the potentiometer.

4.3. Determination of thermodynamic functions of the reaction

4.3.1. Switch the potentiometer to the "off" position and connect all parts of the apparatus as shown in Fig. 2-12-3.

4.3.2. Switch on the super thermostat bath, and adjust the temperature to a specified value, 298K, for example. After a well thermostatic state of the cell is reached, measure the EMF of the cell using the method as described above.

4.3.3. Increase temperature of the water bath for about 5℃ each time, wait for the system to be stable, and then measure the EMF of the cell at different temperatures until five to six data of EMF are obtained.

5. Data recording and processing

5.1. Fill table 2-12-1 and table 2-12-2 with the data measured.

5.2. Calculate pH of the tested solution using equation 2-12-5.

5.3. Plot E versus T, and make linear regression to calculate the temperature coefficient of EMF, $\left(\frac{\partial E}{\partial T}\right)_p$.

5.4. According to the regression equation, calculate E, $\Delta_r G_m$, $\Delta_r H_m$, and $\Delta_r S_m$ at 298K.

6. Questions

6.1. What phenomena will occur if the connection of positive pole and negative pole is mistaken? What if circuit doesn't connect?

6.2. Why should the zinc electrode be prepared into an amalgam electrode? Will it have any influence in zinc electrode potential?

6.3. Why the compensation method should be used to measure the electromotive force?

6.4. How to determine the equilibrium constant for a cell reaction by EMF measurement?

6.5. What are the important roles of working cell and standard cell in the experiment?

7. Pharmaceutical applications

pH determination in pharmaceutical research is of great importance. Since the decomposition processes of many drugs in formulation can be described as oxidation-reduction (redox) reactions and H^+ ions participate in these reactions. So the decomposition is pH dependent. For example, apomorphine is more stable in acidic formulations and rapidly decomposes in basic or neutral solutions. Usually pH must be controlled in liquid formulations to inhibit redox reactions.

实验十三 旋光法测定蔗糖转化反应的速率常数

（一）实验目的

1. 掌握一级反应动力学的特点和研究方法。
2. 测定一级反应的速率常数，并计算反应的半衰期。
3. 了解旋光仪的基本原理，掌握其使用方法。
4. 了解反应动力学的药学应用。

（二）实验原理

在有氢离子存在时，蔗糖水溶液将水解生成葡萄糖及果糖，反应式为

$$C_{12}H_{22}O_{11}(\text{蔗糖})+H_2O \xrightarrow{H^+} C_6H_{12}O_6(\text{葡萄糖})+C_6H_{12}O_6(\text{果糖})$$

实验证明，该反应的速率与蔗糖、水及催化剂 H^+ 的浓度均有关。由于水溶液中，溶剂水的浓度基本不变，而 H^+ 是催化剂，其浓度也保持不变，这时反应速率只与蔗糖浓度有关，可视为准一级反应，其动力学方程为

$$\ln c=\ln c_0-kt \qquad \text{式(2-13-1)}$$

式中 k 为反应速率常数；c 为 t 时间的反应物浓度；c_0 为反应物的初始浓度。

当 $c=\frac{c_0}{2}$ 时，反应所需的时间称为反应的半衰期，用 $t_{1/2}$ 表示。由式(2-13-1)可得

$$t_{1/2}=\frac{\ln 2}{k}=\frac{0.693}{k} \qquad \text{式(2-13-2)}$$

只要测得不同时刻反应物和产物的浓度，就可由式(2-13-1)和式(2-13-2)求得反应的速率常数和半衰期。

本实验中所用的蔗糖及水解产物均为旋光性物质，但它们的旋光能力不同，故可以利用系统在反应过程中旋光度的变化来跟踪反应的进程。

溶液的旋光度与溶液中所含旋光物质的旋光能力、溶剂的性质、溶液浓度、液层厚度、光源波长及温度等因素有关。为了比较各种物质的旋光能力，引入比旋光度的概念，可用下式表示，即

$$[\alpha]_D^t=\frac{\alpha}{lc} \qquad \text{式(2-13-3)}$$

式中 t 为实验温度；D 为光源波长；α 为旋光度；l 为液层厚度；c 为浓度。式(2-13-3)表明，比旋光度为给定光源波长和特定温度下，单位溶液浓度和单位液层厚度下的旋光度，是物质的

特性常数。

当其他条件不变时，旋光度 α 与浓度 c 成正比，即

$$\alpha = Kc \qquad \text{式(2-13-4)}$$

式中 K 为比例常数，与物质旋光能力、溶剂性质、液层厚度、光源波长、温度等因素有关。

在蔗糖的水解反应中，反应物蔗糖是右旋性物质，其比旋光度$[\alpha]_D^{20}=66.6°$。产物中葡萄糖也是右旋性物质，其比旋光度$[\alpha]_D^{20}=52.5°$，而果糖则是左旋性物质，其比旋光度$[\alpha]_D^{20}=-91.9°$。由于各旋光物质的旋光度具有加和性，因此，随着水解反应的进行，溶液的旋光性将逐渐由右旋变为左旋。最后，当蔗糖完全转化为产物时，左旋角度达到最大值。

若反应时间为 0、t 和∞时，溶液的旋光度分别用 α_0、α_t 和 α_∞ 表示，则当蔗糖未转化时，系统的旋光度 α_0 为

$$\alpha_0 = K_{反}\ c_0 \qquad \text{式(2-13-5)}$$

当蔗糖已完全转化时，系统的旋光度 α_∞ 为

$$\alpha_\infty = K_{生}\ c_0 \qquad \text{式(2-13-6)}$$

式中的 $K_{反}$ 和 $K_{生}$ 分别为反应物和生成物的比例常数。这样，当反应进行到任意时刻，系统的旋光度 α_t 为

$$\alpha_t = K_{反}\ c + K_{生}(c_0 - c) \qquad \text{式(2-13-7)}$$

联立式(2-13-5)、(2-13-6)和(2-13-7)，可以得到

$$c_0 = \frac{\alpha_0 - \alpha_\infty}{K_{反} - K_{生}} = K'(\alpha_0 - \alpha_\infty) \qquad \text{式(2-13-8)}$$

$$c = \frac{\alpha_t - \alpha_\infty}{K_{反} - K_{生}} = K'(\alpha_t - \alpha_\infty) \qquad \text{式(2-13-9)}$$

将式(2-13-8)和式(2-13-9)代入式(2-13-1)，即得

$$\ln(\alpha_t - \alpha_\infty) = -kt + \ln(\alpha_0 - \alpha_\infty) \qquad \text{式(2-13-10)}$$

由此可见，以 $\ln(\alpha_t - \alpha_\infty)$ 对 t 作图为一直线，由该直线的斜率可求得反应速率常数 k，进而可求得半衰期 $t_{1/2}$。

若测得不同温度下的速率常数，根据阿伦尼乌斯方程，可求取反应的活化能 E_a。

$$\ln k = -\frac{E_a}{RT} + \ln A \qquad \text{式(2-13-11)}$$

即以 $\ln k$ 对 $1/T$ 作图，由所得直线的斜率可求出反应的活化能。

（三）仪器与药品

1. 仪器　旋光仪（如 WZZ-2 型自动旋光仪或 301 型旋光仪），恒温槽/超级恒温槽（当使用带恒温水隔套的旋光管时），台秤，秒表，100ml 烧杯，25ml 移液管，100ml 带塞锥形瓶。

2. 药品　蔗糖（分析纯），3mol/L HCl 溶液。

（四）实验步骤

1. 熟悉旋光仪的构造，学习使用方法，了解注意事项。

2. 旋光仪零点的校正　在旋光管夹套中通入 30℃ 恒温水，管内装入蒸馏水，恒温后，测定旋光度，重复测量三次，取其平均值，作为旋光仪的零点。

3. 用台秤称取 20g 蔗糖，加入 100ml 蒸馏水配成 20% 的溶液，若溶液混浊则需过滤。用移液管移取 25ml 蔗糖溶液于干燥的 100ml 带塞锥形瓶中，移取 25ml 3mol/L HCl 溶液于另一锥形瓶中，两者分别放入 30℃ 恒温槽中恒温。

也可配制10%蔗糖溶液和按1∶4稀释的盐酸溶液进行实验。

4. 迅速将HCl溶液倒入蔗糖溶液中,计时开始。为了使两者完全定量混合,将溶液倒回装HCl的锥形瓶中,摇匀,再倒回原来瓶中,来回倒三次。用少量混合液润洗旋光管2次,然后将混合液装满旋光管,进行α_t的测定。从计时开始,每隔3分钟测一次旋光度,测定6次,继而每隔5分钟测一次,测定3次。

使用普通旋光管时,应将旋光管置于恒温槽中,测定前迅速取出,两头擦净后进行测定。测定结束后,再迅速放回到恒温槽中。

5. α_∞的测定 将步骤4剩余的混合液放入50~60℃的恒温水浴中,反应60分钟后冷却至实验温度,测定其旋光度,此值即为α_∞。

6. 根据需要,还可选做以下实验。

(1) 催化剂的用量对反应速率的影响:用蒸馏水将3mol/L HCl稀释成1.5mol/L,重复步骤4、5,测定α_t和α_∞,计算速率常数。

(2) 温度对反应速率的影响:分别在不同温度(如25℃、30℃和35℃)下,使用相同浓度的催化剂,重复步骤4、5,测定α_t和α_∞,计算各温度下的速率常数和反应的活化能。不同温度测定时,取样时间间隔和反应总时间应作适当调整。

7. 注意事项

(1) 装上液体后的旋光管光路中不应存有气泡,旋紧旋光管两端的旋光片时既要防止过松引起液体渗漏,又要防止过紧造成用力过大而压碎玻片。

(2) 操作时应特别注意避免酸液滴漏到仪器上腐蚀仪器,实验结束后必须将旋光管洗净。

(3) 旋光仪中的钠光灯不宜长时间开启,测量间隔较长时应熄灭,以免损坏及温度对α_t的测定产生影响。

(4) 水浴温度不可太高,否则将产生副反应,溶液颜色变黄。同时在恒温过程中避免溶液蒸发影响浓度,以致造成α_∞的偏差。

(五) 数据处理

1. 按表2-13-1记录实验数据。

表2-13-1 不同时刻反应体系的旋光度数据

Table 2-13-1 Optical Rotations of the Reaction System at Different Time

α_0=________;c(HCl)=________ mol/L;c(蔗糖,sucrose)=________%;

T:________℃;p:________ kPa

t/min	α_t	$\alpha_t-\alpha_\infty$	$\ln(\alpha_t-\alpha_\infty)$
0			
…			
∞			

2. 以$\ln(\alpha_t-\alpha_\infty)$对$t$作图,由直线的斜率求出反应速率常数$k$;计算蔗糖转化反应的半衰期。

3. 比较催化剂浓度对反应速率常数k及α_∞的影响。

4. 根据阿伦尼乌斯方程,计算反应活化能E_a。

（六）思考题

1. 实验中，为什么用蒸馏水来校正旋光仪的零点？在蔗糖转化反应过程中，所测的旋光度 α_t 是否需要零点校正？为什么？

2. 如何判断某一旋光物质是左旋还是右旋？

3. 蔗糖溶液为什么可粗略配制？配制反应液时为什么要用移液管取蔗糖和盐酸溶液？

4. 蔗糖的转化速率和哪些因素有关？

（七）药学应用

化学动力学在药学领域有着广泛的应用。如药物在体内的吸收、分布、代谢以及排泄等过程都涉及动力学问题，许多药物的吸收和代谢过程就符合一级反应规律。另外药物制剂的稳定性以及有效期的预测也都涉及化学动力学的知识。药物在一般的保存条件下，会发生分解等化学反应，导致有效成分的含量降低，或者分解产物（杂质）的浓度升高。那么，有效成分的含量降低到达不到疗效浓度需要多长时间，就需要通过化学动力学的实验确定，从而实现对稳定性、有效期以及生产条件的监控。

（八）其他实验系统——硫酸链霉素在碱性溶液中水解

硫酸链霉素在碱性溶液中水解为链霉胍及链霉双糖胺，链霉胍部分分子重排为麦芽酚。

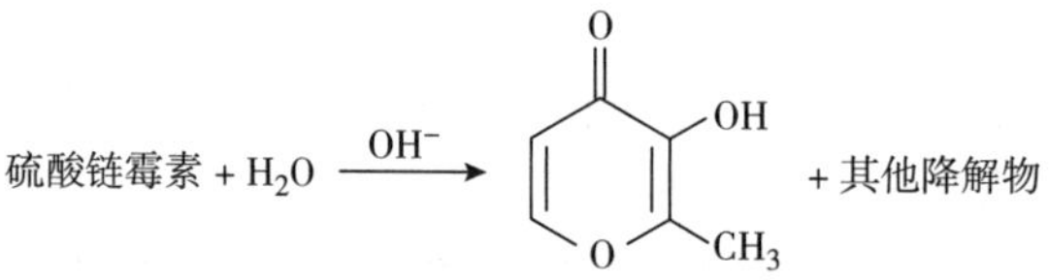

硫酸链霉素 + H_2O $\xrightarrow{OH^-}$ （麦芽酚：O、OH、O、CH_3）+ 其他降解物

该反应为假一级反应，反应速率服从一级反应动力学方程式，即

$$\ln(c_0-c)=-kt+\ln c_0 \qquad \text{式(2-13-12)}$$

式中 c_0 为链霉素的初浓度；c 为 t 时刻麦芽酚的浓度；k 为水解速率常数。

硫酸链霉素水溶液在碱性条件下水解时，定量产生的麦芽酚可与铁离子作用，生成紫红色络合物，该络合物在 520nm 波长处有最大吸收，由溶液的吸光度可以测得不同时刻硫酸链霉素的浓度。

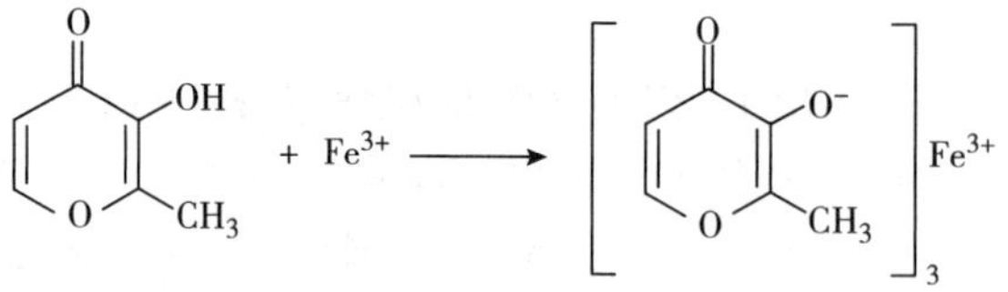

（麦芽酚：O、OH、O、CH_3）+ Fe^{3+} ⟶ [（O、O^-、O、CH_3）]$_3$ Fe^{3+}

硫酸链霉素水溶液的初浓度 c_0 与此溶液中硫酸链霉素全部水解时测得的吸收度 A_∞ 成正比，即 $c_0 \propto A_\infty$；t 时刻麦芽酚的浓度 c 与该时间测得的吸收度 A_t 成正比，即 $c \propto A_t$，而 t 时刻硫酸链霉素（c_0-c）的浓度正比于（$A_\infty-A_t$）。根据式（2-13-12），可以得到

$$\ln(A_\infty-A_t)=-kt+\ln A_\infty \qquad \text{式(2-13-13)}$$

以 $\ln(A_\infty-A_t)$ 对 t 作图为一直线，由直线的斜率计算反应的速率常数 k。

（九）仪器使用说明

旋光仪是测定物质旋光度的仪器。通过对样品旋光度的测定，可以分析确定物质的浓度、含量及纯度。

1. 301 型旋光仪介绍

（1）仪器的构造原理：旋光仪的主要元件是两块尼柯尔棱镜，用于产生平面偏振光的棱镜称为起偏镜，另一棱镜称为检偏镜。通过调节检偏镜，能使透过的光线强度在最强和零之间变化。如果在起偏镜与检偏镜之间放有旋光性物质，则由于物质的旋光作用，使来自起偏镜的光的偏振面改变了某一角度，只有检偏镜也旋转同样的角度，才能补偿旋光线改变的角度，使透过的光的强度与原来相同。旋光仪就是根据这种原理设计的，如图 2-13-1 所示。

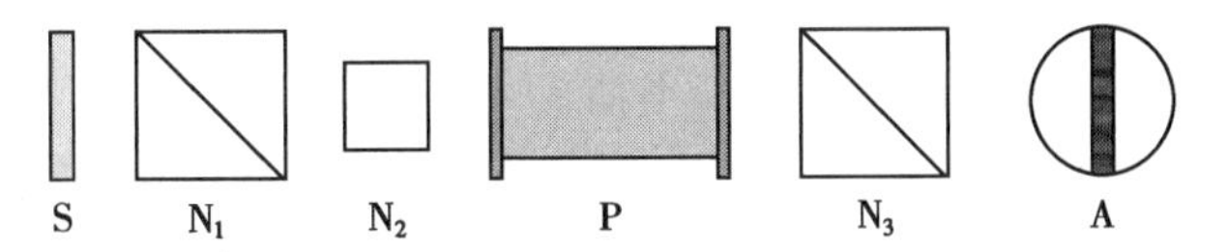

S：钠光光源（sodium light source）；N_1：起偏镜（polarizer）；N_2：石英片（quartz plate）；P：旋光管（polarimeter tube）；N_3：检偏镜（analyzer）；A：目镜视野（vision field observed with eyepiece）

图 2-13-1 旋光仪构造示意图

Fig. 2-13-1 Schematic Diagram of Polarimeter Construction

通过检偏镜用肉眼判断偏振光通过旋光物质前后的强度是否相同是十分困难的，这样会产生较大的误差。为此设计了一种在视野中分出三分视界的装置，如图 2-13-2 所示。其原理是在起偏镜后放置一块狭长的石英片，由起偏镜透过来的偏振光通过石英片时，由于石英片的旋光性，使偏振旋转了一个角度 Φ，通过镜前观察。

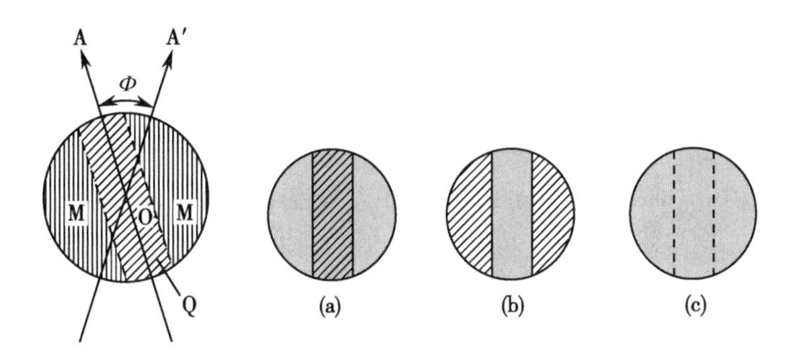

图 2-13-2 三分视野示意图

Fig. 2-13-2 Schematic Diagram of Trisection Visual Fields

A 是通过起偏镜的偏振光的振动方向，A′是又通过石英片旋转一个角度后的振动方向，此两偏振方向的夹角 Φ 称为半暗角（$\Phi=2°\sim3°$）。如果检偏镜的偏振面与起偏镜的偏振面平行（即在 A 的方向时），视野中观察到的是中间狭长部分较暗而两旁较亮，如图 2-13-2（a）所示。如果旋转检偏镜使透射光的偏振面与 A′平行时，在视野中将观察到中间狭长部分较明亮，而两旁较暗，如图 2-13-2（b）所示。当检偏镜的偏振面处于 $\Phi/2$ 时，中间和两边的光偏振面都被旋转了 $\Phi/2$，故视野呈微暗状态，且三分视野内的暗度是相同的，如图 2-13-2（c）所示，将这一位置作为仪器的零点，在每次测定时，调节检偏镜使三分视界的暗度相同，然后读数。

为便于操作，301 型旋光仪的光学系统以倾斜 20°安装在基座上。光源采用 20W 钠光灯（$\lambda=5\ 893Å$）。钠光灯的限流器安装在基座的底部。检偏器与读数圆盘共同连接转动手轮，转动手轮可调整三分视野。仪器采用双游标读数，以消除刻度盘偏心差。读数盘分为 180 小格，每格 1°，游标分 20 格，读数精度为 0.05°（图 2-13-3）。游标窗前方安装有两块 4 倍放大镜，供读数使用。

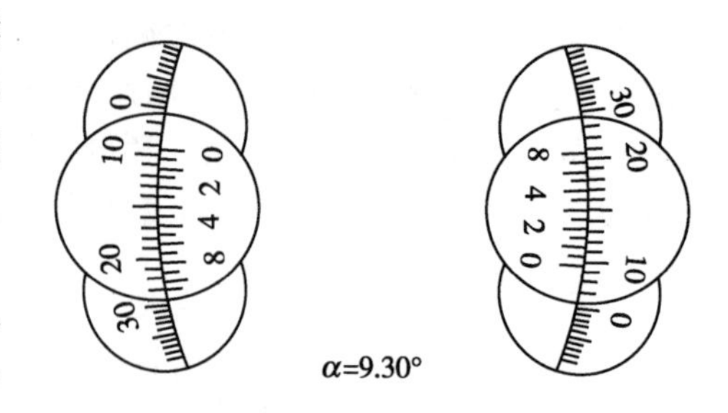

图 2-13-3 301 型旋光仪读数盘
Fig. 2-13-3 Dial indicators of 301 type polarimeter

（2）使用方法：首先打开钠光灯，稍等几分钟，待光源稳定后，从目镜中观察视野，如不清楚可调节目镜焦距。选用合适的样品管并洗净，充满蒸馏水（应无气泡），放入旋光仪的样品管槽中，调节检偏镜的角度使三分视野消失，读出刻度盘上的刻度并将此角度作为旋光仪的零点。零点确定后，将样品管中蒸馏水换为待测溶液，按同样方法测定，此时刻度盘上的读数与零点读数之差即为该样品的旋光度。

（3）使用注意事项：旋光仪在使用时，需通电预热几分钟，但钠光灯使用时间不宜过长。旋光仪是比较精密的光学仪器，使用时，仪器金属部分切忌污染酸碱，防止腐蚀。光学镜片部分不能与硬物接触，以免损坏镜片。不能随便拆卸仪器，以免影响精度。

2. WZZ-2 型自动旋光仪介绍

（1）仪器的构造原理：WZZ-2 型自动旋光仪采用光电检测自动平衡原理，进行自动测量，测量结果由数字显示。

WZZ-2 型自动旋光仪的测量原理如图 2-13-4 所示。仪器采用 20W 钠光灯做光源，由小孔光栏和物镜组成一个简单的点光源平行光管，平行光经偏振镜 A（偏振轴为 OO）变为平面偏振光，当偏振光经过有法拉第效应的磁旋线圈时，其振动平面上产生 50Hz 的 β 角摆动，光线经过偏振镜 B（偏振轴为 PP）投射到光电管上，产生交变的电讯号。

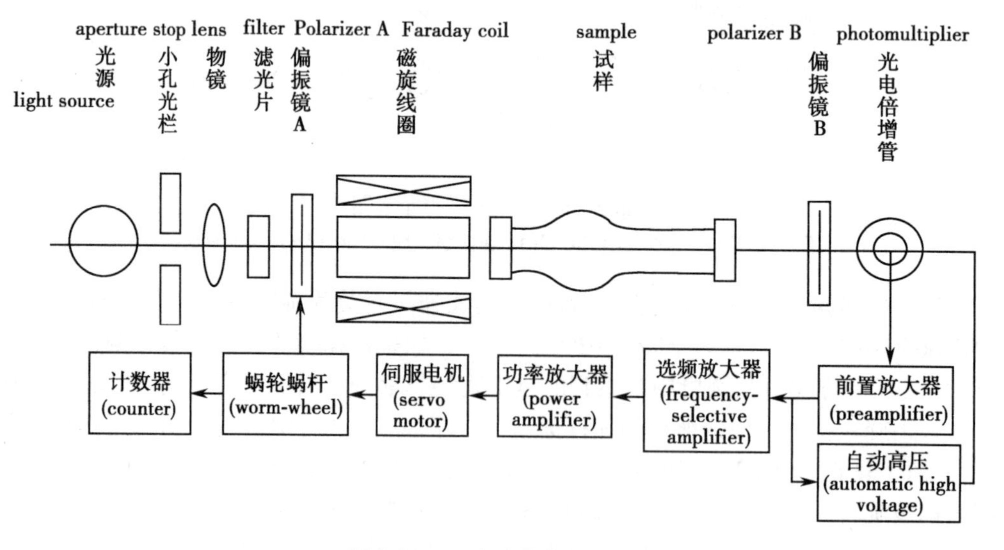

图 2-13-4 自动旋光仪原理
Fig. 2-13-4 Principle of Automatic Polarimeter

仪器以两偏振镜光轴正交时（即 OO ⊥ PP）作为光学零点，此时，$\alpha = 0°$。当偏振光通过旋光性物质时，偏振光的振动面与偏振镜 B 的偏振轴不垂直，光电检测器便能检测到 50Hz

的光电讯号,该讯号能使工作频率为 50Hz 的伺服电机转动,并通过蜗轮蜗杆将偏振镜 A 转过一个 α 角度,此 α 角度就是被测试样的旋光度,并在旋光仪的数显窗显示。

(2) 使用方法

1) 将仪器电源插头插入 220V 交流电源,打开电源开关,这时钠光灯应启亮,预热 5 分钟,使钠光灯发光稳定。

2) 打开光源开关,如光源开关扳上后,钠光灯熄灭,则再将光源开关上下重复扳动 1 到 2 次,使钠光灯在直流下点亮,为正常。

3) 打开测量开关,这时数码管应有数字显示。

4) 将装有蒸馏水或其他空白溶剂的试管放入样品室,盖上箱盖,待示数稳定后,按清零按钮。

5) 取出空白溶剂的试管,将待测样品注入试管,按相同的位置和方向放入样品室内,盖好箱盖。仪器数显窗将显示出该样品的旋光度。

6) 逐次揿下复测按钮,重复读几次数,取平均值作为样品的测定结果。

7) 如果样品超过测量范围,仪器在±4 处来回振荡。此时,取出试管,仪器即自动转回零位。

8) 仪器使用完毕后,应依次关闭测量、光源、电源开关。

(3) 注意事项

1) 仪器应放在干燥通风处,防止潮气侵蚀,尽可能在 20℃ 的工作环境中使用仪器,搬动仪器应小心轻放,避免震动。

2) 在调零或测量时,试管中不能有气泡,若有气泡,应先让气泡浮在凸颈处;如果通光面两端有雾状水滴,应用软布揩干。试管螺帽不宜旋得太紧,以免产生应力,影响读数。试管安放时应注意标记的位置和方向。

3) 钠灯在直流供电系统出现故障不能使用时,仪器也可在钠灯交流供电的情况下测试,但仪器的性能可能略有降低。

Experiment 13 Determination of Rate Constant for the Conversion of Sucrose by Polarimetric Method

1. Objective

1.1. To master the characteristics and study method of the first order kinetics.

1.2. To determine the rate constant and half life of the reaction.

1.3. To understand the principle and operation method of the polarimeter.

1.4. To learn the pharmaceutical application of chemical kinetics.

2. Principle

The hydrolytic reaction of sucrose is as follows

$$C_{12}H_{22}O_{11}(\text{sucrose}) + H_2O \xrightarrow{H^+} C_6H_{12}O_6(\text{glucose}) + C_6H_{12}O_6(\text{fructose})$$

It is known that the reaction rate depends on concentrations of sucrose, water and hydrogen ions (catalyst). In this experiment, the concentrations of water and H^+ ions are constant, then the hydrolysis of sucrose follows pseudo-first order kinetics

$$\ln c = \ln c_0 - kt \tag{2-13-1}$$

where k is the rate constant, c and c_0 are concentrations of sucrose at time t and initial time respectively. Equation(2-13-1) indicates that the rate constant k may be calculated from a plot of $\ln c_A$ against t.

A common parameter used to describe the kinetics of rate processes is the half life $t_{1/2}$. This is simply the time required for half of the reactant to disappear and the $t_{1/2}$ for a first order reaction can be expressed as

$$t_{1/2}=\frac{\ln 2}{k}=\frac{0.693}{k} \tag{2-13-2}$$

In this experiment the rate of reaction between sucrose and water catalyzed by H^+ ions are followed by measuring the optical rotation of the solution with a polarimeter since sucrose, glucose and fructose are optically active substances.

The optical rotation of a solution is affected by optically active species, solute property, concentration of solution, path length, wavelength of light source and temperature etc. The specific rotation of a pure substance is an intrinsic property of that substance at a given wavelength and temperature, and can be used to compare the optical rotatory power among different substances. The specific rotation is defined as

$$[\alpha]_D^t=\frac{\alpha}{lc} \tag{2-13-3}$$

where t is the experimental temperature, D is the wavelength of the light source, α is the optical rotation, l is the path length, and c is the concentration. If all the other variables are kept constant, the optical rotation of a solution is directly proportional to the concentration of the optically active species in solution, that is

$$\alpha=Kc \tag{2-13-4}$$

where K is a constant which is related to optically active species, solute, pass length, wavelength and temperature etc.

Sucrose is dextrorotatory, the specific rotation being 66.6°, glucose is also dextrorotatory, $[\alpha]_D^{20}=52.5°$, but fructose is levorotatory with a large negative specific rotation, $[\alpha]_D^{20}=-91.9°$. The resulting mixture of glucose and fructose is slightly levorotatory because the levorotatory fructose has a greater molar rotation than the dextrorotatory glucose. Hence, the dextrorotatory angle decreases as the reaction proceeds. After a zero rotation is obtained, the system becomes levorotatory and reaches a maximum value at the end of the reaction.

Assuming that the α_t, α_∞ and α_0 are optical rotations at time t, t_∞ and t_0, respectively, where α_t is contributed by the remaining sucrose and the products, α_∞ is contributed by glucose and fructose, and α_0 is contributed by sucrose, we have

$$\alpha_0=K_{\text{Sucrose}}c_0 \tag{2-13-5}$$

$$\alpha_\infty=(K_{\text{Glucose}}+K_{\text{Fructose}})c_0 \tag{2-13-6}$$

$$\alpha_t=K_{\text{Sucrose}}c+(K_{\text{Glucose}}+K_{\text{Fructose}})(c_0-c) \tag{2-13-7}$$

From equations(2-13-5), (2-13-6) and (2-13-7), we obtain

$$c_0=\frac{\alpha_0-\alpha_\infty}{K_{\text{Sucrose}}-(K_{\text{Glucose}}+K_{\text{Fructose}})}=K'(\alpha_0-\alpha_\infty) \tag{2-13-8}$$

$$c = \frac{\alpha_t - \alpha_\infty}{K_{\text{Sucrose}} - (K_{\text{Glucose}} + K_{\text{Fructose}})} = K'(\alpha_t - \alpha_\infty) \tag{2-13-9}$$

Substituting c_0 and c from equations(2-13-8) and(2-13-9) into equation(2-13-1), we have

$$\ln(\alpha_t - \alpha_\infty) = -kt + \ln(\alpha_0 - \alpha_\infty) \tag{2-13-10}$$

Equation(2-13-10) indicates that the rate constant of the reaction may be found from the slope of a plot of $\ln(\alpha_t - \alpha_\infty)$ against t. Furthermore, $t_{1/2}$ can be calculated from equation(2-13-2).

For determination of activation energy of the reaction, E_a, the rate constants at different temperatures should be measured. According to Arrhenius equation, we have

$$\ln k = -\frac{E_a}{RT} + \ln A \tag{2-13-11}$$

Plot $\ln k$ against $1/T$, the activation energy of the reaction can be calculated from the slope.

3. Apparatus and chemicals

3.1. Apparatus

Polarimeter(WZZ-2 type automatic or 301 type), thermostat/super thermostat (when water-jacketed polarimeter tube is used), platform balance, stopwatch, 100ml beaker, 25ml pipettes, 100ml conical flasks with stopper.

3.2. Chemicals

Sucrose(AR), 3mol/L HCl solution.

4. Procedures

4.1. Be familiar with the construction of polarimeter, read the operation manual carefully before use.

4.2. Calibration of the polarimeter

Determine the optical rotation of the distilled water for three times at 30℃. Average the rotations and take it as the zero reading of the polarimeter.

4.3. Dissolve 20g of pure sucrose in distilled water and dilute to 100ml to prepare a 20% solution(filtered, if necessary, to give a clear solution). Adjust the temperature of thermostat to 30℃. Pipette out 25ml sucrose solution into a 100ml conical flask and 25ml of 3mol/L HCl solution into another 100ml conical flask. Put the flasks into the thermostat.

A 10% sucrose solution and four times diluted HCl solution can also be prepared to run the experiment.

4.4. Determination of α_t

Add HCl into the sucrose solution quickly, and start the stopwatch simultaneously to record the reaction time. Mix the solution three or four times by pouring the mixed solution back into the flask in which holding the HCl solution previously and then pouring it into the flask holding the mixture. Use small portions of the mixed solution to rinse out the polarimeter tube for two times. Then fill the tube rapidly with the solution, and determine the optical rotation α_t at a time interval of 3min for 6 times. Since the reaction slows down with the lapse of time, the subsequent readings are recorded at a time interval of 5min for 3 times.

When a common polarimeter tube is used in the experiment, it should be put in the thermostatic bath. Take out the tube just before measuring. When the measurement is finished, put the tube back to the bath immediately.

4.5. Determination of α_∞

Place rest of the reaction solution prepared in procedure 4.4 in a 50～60℃ water bath for 60min to allow a complete reaction. Cool it to the experimental temperature, and measure the optical rotation α_∞.

4.6. If it is necessary, continue the experiment as follows.

4.6.1. The effect of catalyst on the reaction rate: Dilute the 3mol/L solution of HCl to 1.5mol/L with distilled water, and measure the α_t and α_∞ again according to procedures 4.4 and 4.5 for calculating the rate constant.

4.6.2. The effect of temperature on reaction rate: Measure α_t and α_∞ at different temperatures (such as 25℃, 30℃ and 35℃), calculate the rate constants at these temperatures, and then calculate activation energy of the reaction. Notice that the interval and the total reaction time are better to be adjusted when experiment at different temperatures.

4.7. Attentions

4.7.1. Closely attach the glass plate to the polarimeter tube, and keep no air bubbles in the tube. When the glass plate is screwed up, do not overexert.

4.7.2. Spills of acid on the polarimeter are not allowed because they are corrosive. Clean and dry the polarimeter tube once the determination is finished.

4.7.3. Do not turn on the sodium lamp for a long time to prevent damage of the lamp and the influence of temperature in readings as well.

4.7.4. Too high of the temperature of the water bath will cause a side reaction resulting in a yellowish solution. Evaporation of the solution in thermostat will change the concentration of the solution and affect the result of α_∞.

5. Data recording and processing

5.1. Fill table 2-13-1 with the date measured and calculated

5.2. Plot $\ln(\alpha_t - \alpha_\infty)$ versus t. Calculate the reaction rate constant k from the slope of the straight line. Calculate half life $t_{1/2}$ of the reaction.

5.3. Discuss the influence of concentration of HCl in k and α_∞.

5.4. Calculate activation energy of the reaction from Arrhenius equation.

6. Questions

6.1. Why is the distilled water employed to calibrate the zero point of the polarimeter in the experiment? Is it necessary to make a zero-point calibration for α_t determined during sucrose conversion? Why?

6.2. How to judge a dextrorotatory solution or levorotatory solution?

6.3. Why sucrose solution can be prepared roughly? Why solutions of sucrose and HCl must be transferred using pipette?

6.4. What factors will influence the conversion rate of sucrose?

7. Pharmaceutical applications

Chemical kinetics has been widely used in pharmaceutical practice. For instance, the processes of absorption, distribution, metabolism and excretion of drugs in the body can be measured using kinetic models. For most drugs, absorption and elimination follow first order kinetics. Besides, the stability and expiration date prediction of drug formulations are dependent on the knowledge of the chemical kinetics of pharmaceuticals. Under normal storage conditions, the drug will undergo chemical reactions such as decomposition, resulting in a decrease in the content of active ingredients or an increase in the concentration of decomposition products (impurities). Then, how long it takes for the content of the active ingredients to decrease to reach the curative concentration needs to be determined by chemical kinetic experiments, so as to realize the monitoring of stability, expiration date and production conditions.

8. Other experimental system: Rate constant of Streptomycin sulfate hydrolysis by absorption intensity method

Streptomycin sulfate is hydrolyzed to streptidine and streptobiosamine in alkaline solution with part of the streptidine molecules being rearranged to maltol.

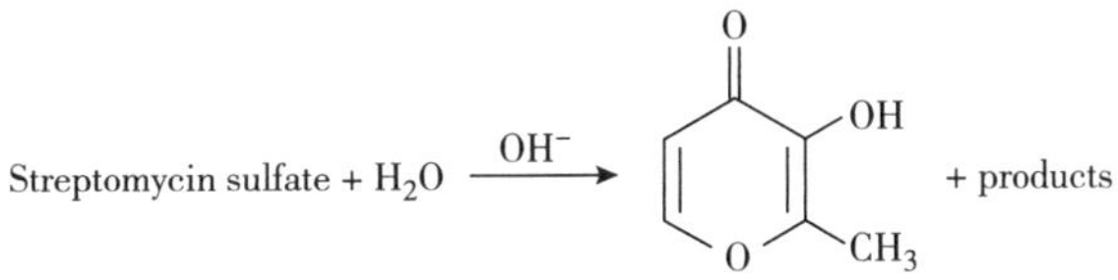

Streptomycin sulfate + H_2O $\xrightarrow{OH^-}$ (maltol: O, OH, O, CH_3) + products

The reaction is regarded as pseudo-first order reaction. The kinetic equation is

$$\ln(c_0-c)=-kt+\ln c_0 \qquad (2\text{-}13\text{-}12)$$

where c_0 is the initial concentration of reactant, c is the concentration of maltol at time t, and k is the rate constant.

It is known that maltol can react with ferri ion to produce a prunosus complex which has a maximum absorption at 520nm. The concentrations of streptomycin sulfate can be determined by measuring the absorbance of the solution at different time.

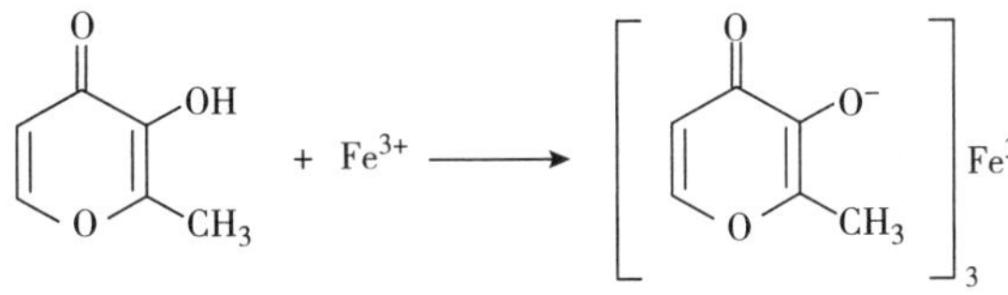

(maltol: O, OH, O, CH_3) + Fe^{3+} $\longrightarrow$ [(O, O^-, O, CH_3)]$_3$ Fe^{3+}

The initial concentration of streptomycin sulfate, c_0, is in direct ratio to the absorbance of the solution when it is completely hydrolyzed, that is $c_0 \propto A_\infty$. The concentration of maltol, c, is in direct ratio to the absorbance of the solution at time t, that is $c \propto A_t$. Therefore, (c_0-c) is in direct ratio to $(A_\infty-A_t)$. From equation (2-13-12), we have

$$\ln(A_\infty-A_t)=-kt+\ln A_\infty \qquad (2\text{-}13\text{-}13)$$

Plot $\ln(A_\infty-A_t)$ versus t, the rate constant k can be calculated from the slope of the line.

9. Operation instructions for polarimeter

The polarimeter is a kind of instrument for measuring the optical rotation of a substance. Through measuring the optical rotation, the polarimeter can be used to analyze the concentration, content, and purity of a substance.

9.1. 301 type polarimeter

9.1.1. Fundamental principle of the instrument: Two Nicol prisms are main components of the polarimeter. The first prism can generate plane-polarized light waves. The second prism can be used as an analyzer. When the transmission plane of analyzer is parallel to polarizer, this polarized light can also pass the second prism completely. While they are perpendicular, the polarized light waves can't pass the second prism at all. Thus, part of polarized light can pass through the second prism when they are crossing. In general, the first prism is fixed unable to move, while the second prism can be turned as the result of the changing light intensity. When a polarized active substance is put between the prisms, the light intensity will be altered. In order to keep the original intensity, the second prism should be turned the same angle as optical rotation. A polarimeter is designed with the principle. An instrument construction is shown in Figure. 2-13-1.

In general, it is difficult to judge the variation of light intensity by naked eye after polarized light pass through the sample if there isn't any reference. Thus, a polarizer with trisection-visual field is invented on which a quartz plate is pasted in the middle of the polarizer. When a polarized light passes through the quartz plate, it will turn Φ angle because of the optical activity of quartz. As shown in Fig. 2-13-2, if the original polarized light waves vibrate in direction of A, that passing through quartz will be A′. The angle between A and A′ is named semidarkness angle Φ which is about 2°~3°. When the polarized plane of analyzer is parallel to A, it can be observed that the middle of visual field is dark and two sides are bright as shown in Fig. 2-13-2(a). When the polarized plane of analyzer is parallel to A′, it can be observed that the middle of the visual field is bright and two sides are dark as shown in Fig. 2-13-2(b). When the polarized plane of analyzer is turn $\Phi/2$ from A, the trisection visual field will be uniform dimness as shown in Fig. 2-13-2(c) which can be taken as the zero point of the instrument. The sample's optical rotation can be read from dial indicator which connected with the analyzer.

For easy operation, the optical system of 301 type polarimeter is installed on the pedestal with 20° obliquity. The light source is 20W sodium lamp ($\lambda = 5\ 893Å$). Both the analyzer and dial indicator are connected to the turning handle that can regulate the trisection visual field. The dial indicator is a round sliding caliper with 180 spaces each representing 1°. The vernier is divided into 20 spaces so that the sensitivity of the polarimeter is 0.05° as shown in Fig. 2-13-3. A 4 times magnifier is installed in front of each dial indicator to make reading clearly.

9.1.2. Operation method

(1) Power on the sodium lamp, and wait for the steady light intensity.

(2) Check up the vision through eyepiece. Adjust the eyepiece focus if the vision is not clear enough.

(3) Fill a clean sample tube with distilled water. Then place it in the sample chamber. Adjust the analyzer to make the trisection visual fields to be uniform dimness. The reading on the dial indicator is taken as zero point.

(4) Remove the distilled water, fill the tube with the sample solution, and determine the optical rotation as described above. The difference of the reading from zero point is the optical rotation of the sample.

9.1.3. Attentions

(1) It is better to preheat the polarimeter for a few minutes before usage. However, the lamp

should not keep lighting for a long time during the experiment. Turn off it whenever not being used.

(2) Polarimeter is a precise instrument. Keep the metal part away from acid or base to avoid being corroded.

(3) Keep the optical lens away from something rigid.

(4) Do not dismantle the polarimeter by the operator himself, otherwise the precision of the instrument will be altered.

9.2. WZZ-2 type automatic polarimeter

9.2.1. Fundamental principle of the instrument: WZZ-2 type automatic polarimeter employs a balance principle of photo electricity to finish the measuring process automatically. Fig.2-13-4 is the illustration of fundamental principle of the instrument. The 20W sodium lamp is used as light source. The parallel light is produced by object lens and diaphragm. Then the parallel light is transformed to plane polarized light through a polarizer A. When it passes through magnetic winding, there is a 50Hz of β angle segmental motion on the plane of vibration, then the light passes through polarizer B and project on the photoelectric tube to produce alternative electric signal.

When the two optical axis of polarizers are crosscut($OO \perp PP$), it is considered as optical zero point, that is $\alpha = 0°$. When polarized light crosses through a sample, the plane of polarizer A will not vertical to the optical axis of polarizer B, and the photoelectric detector will sense a 50Hz signal, which can turn on the servo-generator to make polarizer turn α angle, and the degree of rotation will be shown on the screen.

9.2.2. Operation method

(1) Plug the instrument into the 220V power supply. The sodium lamp lights up. Allow the instrument to warm up for 5min.

(2) Power on the light source switch. If the lamp goes out, the switch should be pulled up and down repeatedly for one or two times to light up the lamp.

(3) Open the measuring switch, and observe whether there is reading on the screen.

(4) Fill the polarimeter tube with distilled water or solvent, place it in the sample chamber, and cover the tank cap. When a stabilized reading is displayed, press the "zero" button.

(5) Take out the polarimeter tube, and fill it with sample solution. Put the tube in the sample chamber, then cover the cap. Now the optical rotation of the sample is displayed on the screen.

(6) Press "measure again" button to repeat the measurement. Average the readings.

(7) If the reading exceeds the measuring range, "±4" will be shown on the screen. Take out the tube, and the reading will get back to zero.

(8) When the measurement is finished, turn off the measuring button, light source and power supply in turn.

9.2.3. Attentions

(1) The instrument should be set in a dry and ventilated position. It is better to service at 20℃. Handle with care and avoid shaking.

(2) Air bubbles are not allowed in the polarimeter tube. If the air bubble is observed, float it

to the protruding site of the tube. The mist on the glass plates should be wiped out with mull. Female screw should not be screwed too tightly to produce a pressure which will affect the measurement result. Pay attention to the site and direction of the tube when it is placed.

(3) Sodium lamp can be used at alternating current if the direct current is not applied, but the efficiency of the instrument may be cut down a little.

实验十四 乙酸乙酯皂化反应速率常数及活化能的测定

(一) 实验目的

1. 用电导法测定乙酸乙酯皂化反应速率常数和半衰期。

2. 了解二级反应的特点，学会用作图法或计算法求取二级反应的速率常数，掌握反应活化能的测定方法。

3. 进一步熟悉和掌握电导率仪的使用方法。

4. 了解化学动力学的药学应用。

(二) 实验原理

乙酸乙酯的皂化反应是典型的二级反应，其反应式为

$$CH_3COOC_2H_5+NaOH \longrightarrow CH_3COONa+C_2H_5OH$$

反应过程中各物质的浓度随时间的变化，可直接通过酸碱滴定求得（化学法），或通过间接测定溶液电导率而求得（物理法），本实验采用后者。为处理方便起见，在设计实验时采用相同的反应物起始浓度 c_0。则反应的动力学方程为

$$\frac{1}{c}-\frac{1}{c_0}=kt \qquad 式(2\text{-}14\text{-}1)$$

式中 k 为反应速率常数。只要测出不同 t 时刻的浓度 c，就可以通过计算得到反应速率常数 k。

若整个反应系统是在稀溶液中进行的，可以认为 CH_3COONa 是全部解离的。因此，反应系统中参与导电的离子有 Na^+、OH^- 和 CH_3COO^-。而 Na^+ 的浓度在反应前后不发生变化，OH^- 的导电能力比 CH_3COO^- 大得多，因此反应进行过程中电导率值将随着 OH^- 被 CH_3COO^- 取代而不断减小。这样，可以用电导率的变化表征浓度的变化。令 κ_0 为反应起始时的电导率，κ_t 为反应进行至 t 时刻的电导率，κ_∞ 为反应完全时的电导率，则有

$$\kappa_0=K_1c_0 \qquad 式(2\text{-}14\text{-}2)$$

$$\kappa_\infty=K_2c_0 \qquad 式(2\text{-}14\text{-}3)$$

$$\kappa_t=K_1c+K_2(c_0-c) \qquad 式(2\text{-}14\text{-}4)$$

将式(2-14-2)和式(2-14-3)代入式(2-14-4)，消去比例常数 K_1 和 K_2，可得浓度 c 与电导率的关系式为

$$c=\frac{\kappa_t-\kappa_\infty}{\kappa_0-\kappa_\infty}c_0 \qquad 式(2\text{-}14\text{-}5)$$

将式(2-14-5)代入式(2-14-1)，得

$$\frac{\kappa_0-\kappa_t}{\kappa_t-\kappa_\infty}=c_0kt \qquad 式(2\text{-}14\text{-}6)$$

因此，只要测定 κ_0、κ_∞ 和不同时刻的 κ_t 值后，以 $\frac{\kappa_0-\kappa_t}{\kappa_t-\kappa_\infty}$ 对 t 作图，由直线斜率可以求出反

应速率常数 k。

对同一电导池而言，电导与电导率成正比。因此，用电导代替电导率，上述关系式仍然成立，即

$$\frac{G_0-G_t}{G_t-G_\infty}=c_0kt \qquad 式(2\text{-}14\text{-}7)$$

在温度变化范围不大时，反应速率常数与温度的关系符合阿伦尼乌斯方程

$$\ln\frac{k_2}{k_1}=-\frac{E_a}{R}\left(\frac{1}{T_2}-\frac{1}{T_1}\right) \qquad 式(2\text{-}14\text{-}8)$$

$$\ln k=-\frac{E_a}{RT}+\ln A \qquad 式(2\text{-}14\text{-}9)$$

通过测定两个不同温度 T_1、T_2 时的反应速率常数 k_1 和 k_2，即可由式(2-14-8)计算出反应的活化能 E_a。若测得多个温度下的速率常数，根据式(2-14-9)，由作图法可求得反应的活化能 E_a和指前因子 A。

（三）仪器和试剂

1. 仪器　恒温槽，DDS-307 型电导率仪，秒表，电导池（或碘瓶，或试管），10ml 移液管。

2. 试剂　0.02mol/L NaOH 溶液，0.02mol/L $CH_3COOC_2H_5$ 溶液，0.01mol/L NaOH 溶液，0.01mol/L CH_3COONa 溶液。

（四）实验步骤

1. 实验装置示意图，见图 2-14-1 所示。

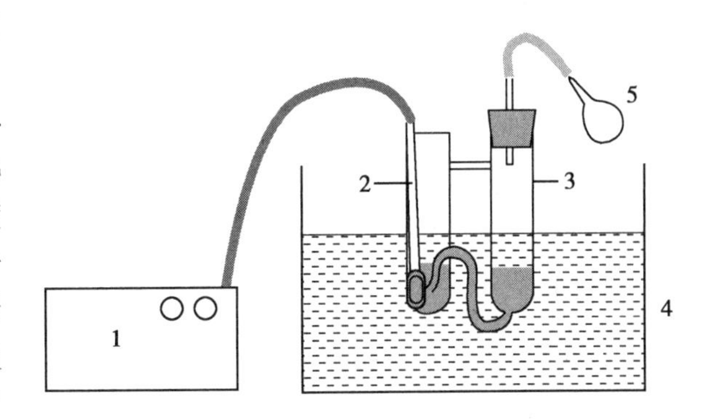

1. 电导率仪（conductometer）；2. 电导电极（conductivity electrode）；3. 双管电导池（double-barrelled conductance cell）；4. 恒温槽（thermostatic bath）；5. 洗耳球（rubber suction bulb）

图 2-14-1　反应速率常数测定示意图

Fig. 2-10-1　Schematic of the Apparatus for Measurement of Reaction Rate Constant

2. 熟悉电导率仪的使用　DDS-307 型电导率仪操作要点：开机预热 10 分钟，将“温度”旋钮调至实验温度（例如 298K）；“校正/测量”开关置于“校正”位置，调节“常数”旋钮使显示数（小数点位置不论）与所使用电极的常数值一致。然后将“校正/测量”开关置于“测量”位置。

3. 调节恒温槽至 25℃。

4. κ_0 的测量　取适量的 0.01mol/L NaOH 溶液于电导池中，插入电导电极，置于恒温槽中恒温 10 分钟。按动“量程/选择”开关选择合适量程（如果数据为 1，表明超出测量范围，应选高挡量程；如读数很小，应选低挡量程），待显示稳定后读数。

5. κ_∞ 的测量　按上述操作方法测量 0.01mol/L CH_3COONa 溶液的电导率 κ_∞。

6. κ_t 的测量　分别用移液管准确吸取 10ml 的 0. 02mol/L NaOH 和 0. 02mol/L $CH_3COOC_2H_5$ 溶液于联通双管电导池中，反应液于分隔的两管中恒温 10 分钟后，用洗耳球将一侧反应物压入另一侧，可压入、抽回反复操作 3 次使其混合均匀，同时开始计时。在第

5、10、15、20、25、30、40 和 50 分钟时分别测其电导率 κ_t。

7. 活化能的测定 选择不同的温度，重复上述步骤，测定不同温度下速率常数，取样时间间隔和反应总时间作适当调整。

8. 注意事项

（1）注意 NaOH 溶液的浓度。

（2）使用电导率仪前要先进行校正。

（3）NaOH 溶液和 $CH_3COOC_2H_5$ 溶液要等体积混合。

（五）数据记录与处理

1. 实验数据记录（表 2-14-1）

表 2-14-1 不同时刻反应系统的电导率值

Table 2-14-1 Conductivity of the Reaction System at Different Time

T:________ K；c_0:________ mol/L；κ_0:________ S/m；κ_∞:________ S/m

t/min	$\kappa_t \times 10^3$/（S/m）	$\frac{\kappa_0-\kappa_t}{\kappa_t-\kappa_\infty}$
0		
5		
10		
15		
20		
25		
30		
40		
50		
∞		

2. 以$\frac{\kappa_0-\kappa_t}{\kappa_t-\kappa_\infty}$对 t 作图，由直线的斜率计算反应速率常数 k，并求算反应的半衰期 $t_{1/2}$。

3. 由不同温度的速率常数值，计算反应活化能。

（六）思考题

1. 为什么 0.01mol/L NaOH 溶液和 0.01mol/L CH_3COONa 溶液的电导率可分别被记为κ_0和κ_∞？

2. 本实验为什么在恒温下进行？改变温度进行上述实验操作，所得的 k 值是否相同？

3. 若 NaOH 溶液和 $CH_3COOC_2H_5$ 溶液起始浓度不等，能否计算 k 值？

（七）药学应用

化学动力学在药学领域有着广泛的应用。在药物合成中，可以通过改变条件，加快目标反应和抑制副反应；在药物的体内代谢中，可测定药物的消除速率以便制订合理的给药方案；在药物制剂中，可预测药物在室温下的反应速率来确定保存期。

（八）其他测量方法

酸碱滴定法：通过酸碱滴定测定不同时刻的 NaOH 浓度，从而直接计算反应速率。

Experiment 14 Determination of Rate Constant and Activation Energy for the Saponification of Ethyl Acetate by Conductometric Method

1. Objective

1.1. To determine the reaction rate constant and half life for the saponification of ethyl acetate by conductometric method.

1.2. To learn the characteristics of second order reaction and to calculate the rate constant of a second order reaction by graphical and calculation method. To master the method of measurement of activation energy of a reaction.

1.3. To learn and master the operation of a conductometer.

1.4. To learn the pharmaceutical application of chemical kinetics.

2. Principle

The kinetics of the saponification of ethyl acetate follows the second-order equation

$$CH_3COOC_2H_5+NaOH \longrightarrow CH_3COONa+C_2H_5OH$$

The concentration variation of the substances over time during the reaction can be determined either directly by acid-base titration method (chemical method) or indirectly by conductometric method (physical method). In this experiment, we use the latter one.

If the initial concentrations of the two reactants are same, we have

$$\frac{1}{c}-\frac{1}{c_0}=kt \tag{2-14-1}$$

where c_0 and c are concentrations of the reactants at $t=0$ and $t=t$, respectively, and k is the rate constant of the reaction.

It can be assumed that CH_3COONa is completely ionized in dilute solution. The ions in the solution accounting for conductance of electricity are Na^+, OH^-, and CH_3COO^-. OH^- has a higher conductive capability than CH_3COO^- Although the concentration of Na^+ remains invariable during the reaction, OH^- is continuously replaced by CH_3COO^-, making the measured conductance of the system falls continuously. Therefore, the variation of concentration can be expressed by the conductance changes.

If κ_0 is the conductivity at $t=0$, κ_t is the conductivity at $t=t$, and κ_∞ is the conductivity at the end of the reaction, we have

$$\kappa_0=K_1c_0 \tag{2-14-2}$$

$$\kappa_\infty=K_2c_0 \tag{2-14-3}$$

$$\kappa_t=K_1c+K_2(c_0-c) \tag{2-14-4}$$

where K is the proportional constant. Substituting equations (2-14-2) and (2-14-3) for equation (2-14-4) gives

$$c=\frac{\kappa_t-\kappa_\infty}{\kappa_0-\kappa_\infty}c_0 \tag{2-14-5}$$

Introducing equation (2-14-5) into equation (2-14-1) yields

$$\frac{\kappa_0-\kappa_t}{\kappa_t-\kappa_\infty}=c_0kt \qquad (2\text{-}14\text{-}6)$$

Plotting $\frac{\kappa_0-\kappa_t}{\kappa_t-\kappa_\infty}$ versus t gives a straight line and the reaction rate constant k can be calculated from the slope of the line.

The conductance is proportional to conductivity for the same conductivity cell. The equation mentioned above is also right if conductivity is replaced by conductance.

$$\frac{G_0-G_t}{G_t-G_\infty}=c_0kt \qquad (2\text{-}14\text{-}7)$$

The temperature dependence of the rate constant has been found to fit the expressions proposed by Arrhenius equation when the temperature variation is within a reasonable range.

$$\ln\frac{k_2}{k_1}=-\frac{E_a}{R}\left(\frac{1}{T_2}-\frac{1}{T_1}\right) \qquad (2\text{-}14\text{-}8)$$

$$\ln k=-\frac{E_a}{RT}+\ln A \qquad (2\text{-}14\text{-}9)$$

where k_1 and k_2 are the rate constants determined at temperatures T_1 and T_2 respectively, E_a is the activation energy of the reaction, A is the pre-exponential factor. E_a and A can be calculated from equation (2-14-8) by determination of the rate constants at two temperatures, or from a plot of $\ln k$ versus $1/T$ as indicated in equation (2-14-9).

3. Apparatus and chemicals

3.1. Apparatus

Thermostat, DDS-307 type digital conductometer, stopwatch, conductance cell (or iodine flask or test tube), 10ml pipette.

3.2. Chemicals

0.02mol/L NaOH solution, 0.02mol/L $CH_3COOC_2H_5$ solution, 0.01mol/L NaOH solution, 0.01mol/L CH_3COONa solution.

4. Procedures

4.1. The schematic of experiment apparatus is shown in Fig. 2-14-1.

4.2. Operation of conductometer

Before operating the conductometer, read the operation instructions carefully.

DDS-307 type digital conductometer: Turn on the power of the conductometer, and warm up the instrument for 10min. Adjust the "temperature" button to match the operation value, for example 298K. Set the "calibration/measurement" switch to "calibration", adjust "constant" button to match the cell constant of the electrode used. Then set the "calibration/measurement" switch to "measurement" mode.

4.3. Turn on the thermostat, and adjust the temperature to 25℃.

4.4. Measurement of κ_0

Fill the conductance cell with appropriate amount of 0.01mol/L NaOH, insert a conductance electrode into it, and keep the system in the thermostat for 10 minutes. Choose the appropriate range and switch the corresponding button to it. After a thermal equilibrium being reached, measure the conductivity of the solution, and take the reading as κ_0.

4.5. Measurement of κ_∞

Repeat the procedure of 4.3 to determine the conductivity of 0.01mol/L CH_3COONa solution, and take the readings as κ_∞.

4.6. Measurement of κ_t

Transfer respectively 10ml of 0.02mol/L NaOH solution and 10ml of 0.02mol/L $CH_3COOC_2H_5$ solution to two separate tubes of the conductance cell. After a thermal equilibrium is reached, press the solution in one tube to the neighbor one by rubber suction bulb, repeat to mix well if necessary. Switch on the stopwatch to record the reaction time. Then measure the conductivity κ_t at predetermined time interval of 5min, 10min, 15min, 20min, 25min, 30min, 40min and 50 min.

4.7. Measurement of the activation energy

Change the temperature of the system to a specified value and repeat the procedures described above to determine the rate constant at the specified temperature.

4.8. Attentions

4.8.1. Pay attention to the concentration of NaOH solution.

4.8.2. Calibration is required before using the conductivity meter.

4.8.3. Mix NaOH solution with the same volume of $CH_3COOC_2H_5$ solution.

5. Data recording and processing

5.1. Fill table 2-14-1 with the data measured and calculated.

5.2. Plot $\frac{\kappa_0-\kappa_t}{\kappa_t-\kappa_\infty}$ versus t. Calculate reaction rate constant k from the slope of the straight line, and then calculate the half life of the reaction.

5.3. Calculate activation energy of the reaction by Arrhenius equation.

6. Questions

6.1. Why can the conductivity of 0.01mol/L NaOH and 0.01mol/L CH_3COONa be considered as κ_0 and κ_∞?

6.2. Why should the reaction be run at constant temperature? Will the rate constant be the same if the temperature is changed?

6.3. How to calculate the rate constant of the reaction if the initial concentrations of $CH_3COOC_2H_5$ and NaOH are different?

7. Pharmaceutical applications

There is a wide application of chemical kinetics in pharmacy. In drug synthesis, we hope to accelerate the target reactions and inhibit the side reactions at the same time. In drug metabolism, we need to determine the drug elimination rate *in vivo* in order to establish a rational dosage regimen. In pharmaceutical preparations, metabolic rate of drugs should be predicted to determine the period of validity.

8. Some other method for rate constant determination

Acid base titration method: Determine the concentration variation of NaOH at certain time interval to calculate the rate constant.

实验十五 碘化钾与过氧化氢反应的速率常数及活化能的测定

（一）实验目的

1. 测定不同温度下碘化钾与过氧化氢反应的速率常数并计算反应的活化能。
2. 掌握反应速率常数与活化能的实验测定方法。
3. 明确活化能的概念及其对反应速率的影响。
4. 了解反应速率常数和活化能的药学应用。

（二）实验原理

阿伦尼乌斯方程是反应速率常数 k 随温度变化关系的经验公式，可表示为

$$k=A\mathrm{e}^{-\frac{E_a}{RT}} \qquad 式(2\text{-}15\text{-}1)$$

式中 E_a为活化能；A 为频率因子或指（数）前因子。

将式（2-15-1）两边取对数，得

$$\ln k=-\frac{E_a}{R}\cdot\frac{1}{T}+\ln A \qquad 式(2\text{-}15\text{-}2)$$

由上式可见，$\ln k$ 对 $1/T$ 作图，可得直线，由直线斜率和截距可分别求出 E_a和 A。

将式（2-15-2）微分，可得

$$\frac{\mathrm{d}\ln k}{\mathrm{d}T}=\frac{E_a}{RT^2} \qquad 式(2\text{-}15\text{-}3)$$

当温度变化范围不大时，将式（2-15-3）分离变量，定积分，可得反应速率常数随温度变化的关系式，即为

$$\ln\frac{k_2}{k_1}=-\frac{E_a}{R}\left(\frac{1}{T_2}-\frac{1}{T_1}\right) \qquad 式(2\text{-}15\text{-}4)$$

式中 k_1 和 k_2 为在温度 T_1 和 T_2 时的反应速率常数。利用上式，在 T_1 和 T_2 时的速率常数 k_1 和 k_2 已知的情况下，可求得反应的活化能 E_a；或已知 E_a和 T_1 时的 k_1，求出任一温度 T_2 下的 k_2。

本实验测定的化学反应为

$$2H^++2I^-+H_2O_2 \longrightarrow 2H_2O+I_2$$

在 KI 的酸性溶液中，加入一定量的淀粉溶液和 $Na_2S_2O_3$ 标准溶液，然后一次加入一定量的 H_2O_2 溶液。在溶液中进行的反应为

$$I^-+H_2O_2 \longrightarrow IO^-+H_2O（慢） \qquad (a)$$

$$2H^++I^-+IO^- \longrightarrow I_2+H_2O（快） \qquad (b)$$

$$I_2+2S_2O_3^{2-} \longrightarrow 2I^-+S_4O_6^{2-}（快） \qquad (c)$$

当溶液中的 $Na_2S_2O_3$ 未消耗完时，溶液是无色的。当溶液中的 $Na_2S_2O_3$ 一经消耗完毕，反应（b）所产生的 I_2 即与溶液中的淀粉作用使溶液变蓝。这时若再加入一定量的 $Na_2S_2O_3$，溶液又变成无色，记录各次蓝色出现的时间，即可得到此时溶液中 H_2O_2 的浓度。

上述反应历程中，反应（a）为控速步骤，故反应速率方程可表示为

$$-\frac{\mathrm{d}c_{H_2O_2}}{\mathrm{d}t}=k'c_{H_2O_2}c_{I^-} \qquad 式(2\text{-}15\text{-}5)$$

随着反应的进行，I^-不断再生。因此，当溶液的体积不变时，c_{I^-}可视作常数。令 $k=k'c_{I^-}$，则得

$$-\frac{dc_{H_2O_2}}{dt}=kc_{H_2O_2} \quad 式(2\text{-}15\text{-}6)$$

积分后可得

$$\ln\frac{c_0}{c_t}=kt \quad 式(2\text{-}15\text{-}7)$$

式中 c_0 为 H_2O_2 在时间 $t=0$ 时的浓度，可用化学分析法直接测出；c_t 为 H_2O_2 在时间 t 时的浓度，可通过记录各次蓝色出现的时间求得。

以 $\ln c_0/c_t$ 对 t 作图，如果得到的是一直线，即可验证上述机制是正确的。

（三）仪器和试剂

1. 仪器　恒温槽，温度计，500ml 容量瓶，150ml 锥形瓶，25ml 酸式滴定管，10ml 量筒，1 000ml三颈瓶，1ml、25ml 移液管，秒表。

2. 试剂　0.4mol/L KI 溶液，3mol/L H_2SO_4，0.2% H_2O_2 溶液，0.05mol/L $Na_2S_2O_3$，0.02mol/L $KMnO_4$ 标准溶液，1% 淀粉溶液。

（四）实验步骤

1. 两次实验温差最好保持在 10℃左右，一次可在室温（如 25℃）下进行，另一次则在较室温略高 10℃的恒温槽中进行，即 35℃。如果室温已经很高，另一次较高温度的实验会因速率过快而造成操作上的困难，这时可通过适当减少 KI 溶液的加入量、减小 H_2O_2 浓度或其加入量来减小反应速率，也可配制冰水浴降低反应温度来减小反应速率。

2. 于 500ml 容量瓶中加入 50ml 0.4mol/L KI 溶液，稀释至约为容量瓶 2/3 的体积后，加入 20ml 3mol/L H_2SO_4 及 5ml 淀粉溶液，再用蒸馏水稀释至刻度，振荡混合均匀。如果 KI 溶液中有游离 I_2 而使淀粉变蓝，则可在稀释前滴加几滴 $Na_2S_2O_3$ 消去蓝色。

3. 如图 2-15-1 所示，将 1 000ml 三颈瓶固定于恒温槽中，装置好温度计、搅拌棒，注意高

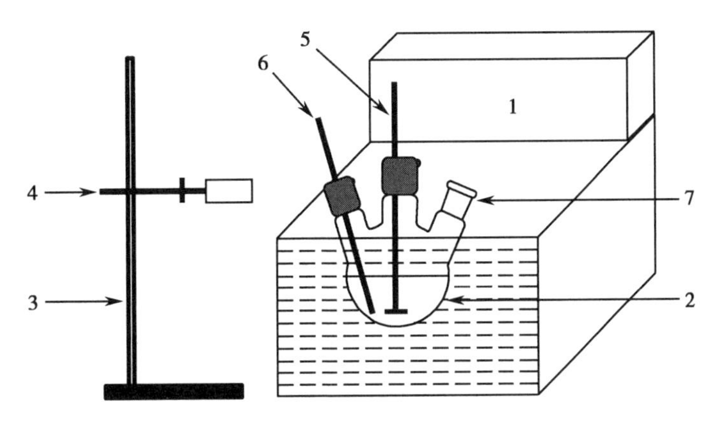

1. 恒温水浴（thermostat）；2. 三颈瓶（three-necked bottle）；3. 铁架台（support stand）；4. 用于三颈瓶固定的固定夹（iron clamp for three-necked bottle fixation）；5.电动搅拌棒（electromotive stirring rod）；6.温度计（thermometer）；7.加样口（liquid feeding）

图 2-15-1 实验装置示意图

Fig. 2-15-1 Schematic Diagram of Experimental Equipment

度调节，不能相互接触，不能接触到容器底部或容器壁。将容量瓶中的溶液倾注入三颈瓶中，开启搅拌器。

4. 待溶液温度与恒温槽温度相差不到1℃时，在溶液内精确移入1ml 0.1mol/L $Na_2S_2O_3$ 溶液，记下溶液温度，立即用移液管加入25ml 0.2% H_2O_2 溶液。当溶液开始出现蓝色时，立即启动秒表，同时精确移入1ml $Na_2S_2O_3$ 溶液。此后每当蓝色出现时，记录时间，同时精确加入1ml $Na_2S_2O_3$ 溶液。当加入 $Na_2S_2O_3$ 溶液的总量已达10ml，记录最后一次蓝色出现时间，并停止实验。

5. 使水浴温度保持与前次实验相差约10℃左右，恒温下重复上述实验。

6. 实验前需用 $KMnO_4$ 溶液标定 H_2O_2 溶液的浓度。方法是：用移液管取25ml H_2O_2 溶液于150ml锥形瓶中，加10ml 3mol/L H_2SO_4，用0.02mol/L $KMnO_4$ 标准溶液滴定至显浅红色为止。

（五）数据记录和处理

1. 按表2-15-1记录实验数据及处理结果。

表2-15-1 数据记录及处理结果

Table 2-15-1 The Data Measured and Calculated

室温（room temperature）：________；大气压（atmospheric pressure）：________

c(0.2% H_2O_2)；c($Na_2S_2O_3$)：________

$V_{Na_2S_2O_3}$/ml		T_1/K		T_2/K	
每次（each time）	累积（accumulated）	t/s	$\ln(c_0/c_t)$	t/s	$\ln(c_0/c_t)$
	k_1 =		k_2 =	E_a =	

2. 设25ml H_2O_2 溶液相当于 $Na_2S_2O_3$ 标准溶液的毫升数为 X，则

$$c_t=\frac{(X-V_{Na_2S_2O_3})\cdot\frac{1}{2}c_{Na_2S_2O_3}}{V+V_{Na_2S_2O_3}} \qquad 式(2\text{-}15\text{-}8)$$

$$c_0=\frac{(X-1)\cdot\frac{1}{2}c_{Na_2S_2O_3}}{V+1} \qquad 式(2\text{-}15\text{-}9)$$

$$\frac{c_0}{c_t}=\frac{(X-1)}{V+1}\cdot\frac{V+V_{Na_2S_2O_3}}{X-V_{Na_2S_2O_3}} \qquad 式(2\text{-}15\text{-}10)$$

式中 $V_{Na_2S_2O_3}$ 为在时间 t 时已加入的 $Na_2S_2O_3$ 溶液总毫升数；$V+V_{Na_2S_2O_3}$ 为溶液在 t 时刻的总体积；$c_{Na_2S_2O_3}$ 为 $Na_2S_2O_3$ 溶液的物质的量浓度。

显然，$V=500+25=525$ml。这里假定了溶液体积具有加和性。c_0 不一定是消耗了1ml

$Na_2S_2O_3$ 溶液后的 H_2O_2 浓度。如果操作中未来得及计时,可在继续加入 2ml、3ml…ml $Na_2S_2O_3$ 后作为 c_0,但这时计算 c_0 的式(2-15-10)中的 1ml 相应地改为 2ml、3ml…ml。

3. 根据式(2-15-7),以 $\ln c_0/c_t$ 为纵坐标,t 为横坐标作图,确定直线的斜率,求出速率常数 k_1、k_2,代入式(2-15-4)计算反应的活化能 E_a。

(六)思考题

1. 在什么条件下才可将本实验中的反应当成一级反应来处理?
2. 本实验中,为什么每当溶液出现蓝色时,需立即加入 $Na_2S_2O_3$ 溶液?
3. 本实验中,KI 溶液的加入量的多少对反应速率是否有影响?为什么?

(七)药学应用

化学反应速率常数和活化能的测定,在药物稳定性预测和作用机制的研究中有广泛的应用。如通过反应速率常数和活化能的测定,研究阿魏酸光致异构化的规律和作用机制,理解其以溶液状态存在的不稳定的原因,从而进一步改进其在抗肿瘤、抗肝纤维化及治疗心、脑血管疾病的作用。

Experiment 15 Determination of Rate Constant and Activation Energy for Reaction between Potassium Iodide and Hydrogen Peroxide

1. Objectives

1.1. To determine the rate constant of reaction between potassium iodide and hydrogen peroxide under different temperatures and calculate its activation energy.

1.2. To master the methods of calculating rate constant and activation energy using experimental data.

1.3. To understand the concept of activation energy and its effect on reaction rate.

1.4. To learn the pharmaceutical applications of rate constant and activation energy.

2. Principles

The Arrhenius equation represents the dependence of rate constant k of a chemical reaction on the temperature T, and can be expressed as

$$k = Ae^{-\frac{E_a}{RT}} \tag{2-15-1}$$

where E_a is the activation energy, and A is the pre-exponential factor or simply the pre-exponential factor.

Taking the natural logarithm of the Arrhenius equation (2-15-1) yields

$$\ln k = -\frac{E_a}{R} \cdot \frac{1}{T} + \ln A \tag{2-15-2}$$

It indicates that a plot of $\ln k$ versus $1/T$ gives a straight line, and the E_a and A can be calculated from the slope and intercept of the line.

Differentiating equation (2-15-2), we have

$$\frac{d\ln k}{dT} = \frac{E_a}{RT^2} \tag{2-15-3}$$

If a moderate temperature change is taken into account, integration of equation (2-15-3) from

T_1 to T_2 gives

$$\ln\frac{k_2}{k_1}=-\frac{E_a}{R}\left(\frac{1}{T_2}-\frac{1}{T_1}\right) \tag{2-15-4}$$

where k_1 and k_2 are the rate constants of reaction at T_1 and T_2 respectively. If k_1 and k_2 at known T_1 and T_2 are given, the activation energy E_a can be calculated using the above equation. If E_a and k_1 at T_1 are given, k_2 at any temperature can be calculated.

In this experiment, the reaction to be studied is

$$2H^+ + 2I^- + H_2O_2 \longrightarrow 2H_2O + I_2$$

When a small and known amount of thiosulfate ($S_2O_3^{2-}$) ions and starch solution to the mixture of peroxide and iodide are added, iodine is produced slowly by the reaction between H_2O_2 and I^- ions and the $S_2O_3^{2-}$ ions immediately react with I_2 as it is produced. The reactions are as follows

$$I^- + H_2O_2 \longrightarrow IO^- + H_2O(\text{slow}) \tag{a}$$

$$2H^+ + I^- + IO \longrightarrow + H_2O(\text{fast}) \tag{b}$$

$$I_2 + 2S_2O_3^{2-} \longrightarrow 2I^- + S_4O_6^{2-}(\text{fast}) \tag{c}$$

As long as excess $S_2O_3^{2-}$ ions present in the solution, the solution is colorless. Once all the $S_2O_3^{2-}$ ions are consumed, I_2 starts to form in the solution and the solution turns blue by the complex of starch-iodine. At this moment, upon addition of some $Na_2S_2O_3$, the solution turns colorless again. By recording the time required for the appearance of blue solution, concentration of H_2O_2 in solution can be calculated.

Since reaction (a) is a rate determining step, the rate equation of the reaction can be written as

$$-\frac{dc_{H_2O_2}}{dt}=k'c_{H_2O_2}c_{I^-} \tag{2-15-5}$$

As the reaction proceeds, consumed I^- ions are regenerated by $S_2O_3^{2-}$ ions. Therefore, c_{I^-} can be regarded as a constant if the volume of the solution keeps unchanged. Assuming $k=k'c_{I^-}$, equation (2-15-5) becomes

$$-\frac{dc_{H_2O_2}}{dt}=kc_{H_2O_2} \tag{2-15-6}$$

Integrating equation (2-15-6) yields

$$\ln\frac{c_0}{c_t}=kt \tag{2-15-7}$$

where c_0 is the concentration of H_2O_2 at t_0 and can be determined by chemical analysis, c_t is the concentration of H_2O_2 at time t and can be determined by recording the time for appearance of bluish solution.

If the plot of $\ln(c_0/c_t)$ versus t yields a straight line, the accuracy of the above mechanism can be proved.

3. Apparatus and Chemicals

3.1. Apparatus

Thermostat, thermometer, 500ml measuring flask, 150ml conical flask, 25ml acid burette, 10ml measuring cylinder, 1 000ml three-necked bottle, 1ml, 25ml pipettes, and stopwatch.

3.2. Chemicals

0.4mol/L KI solution, 3mol/L H_2SO_4, 0.2% H_2O_2 solution, 0.05mol/L $Na_2S_2O_3$ solution, 0.02mol/L $KMnO_4$ standard solution, 1% starch solution.

4. Procedures

4.1. It is better to have 10℃ differences in temperature between two runs of the experiment. For example, when one run is at room temperature, say 25℃, the other is better to run at 35℃. If the room temperature is already higher, operation of the experiment will be difficult due to too much fast reaction rate. In this case, approaches of decreasing the concentration or volume of H_2O_2 will be employed to reduce the reaction rate. Besides, an ice-water bath can also be used for the same purpose.

4.2. Add 50ml of KI solution to a 500ml flask and dilute to 2/3 of its volume. Add 20ml of H_2SO_4 and 5ml of starch solution and dilute with distilled water to volume. Shake it for a thorough mixing. If a blue solution appears due to free I_2 in iodide solution, add a few drops of $Na_2S_2O_3$ to eliminate the bluish appearance before dilution.

4.3. Fix the 1 000ml three-necked bottle as shown in fig 2-15-1. Set the thermometer and electromotive stirrer accordingly. Pay attention to height adjustment and make sure they don't touch each other or touch the bottom or sides of the container. Transfer the solution into the three-necked bottle. Then start the stirrer.

4.4. Wait until a less than 1℃ difference in temperature between the solution and the thermostat has been reached, then add 1ml of $Na_2S_2O_3$ to the solution. Record the temperature of the solution and add 25ml of H_2O_2 instantly. Once the solution turns blue, start the stopwatch and add 1ml of $Na_2S_2O_3$ simultaneously. Repeat the operation until a maximum volume of 10ml of $Na_2S_2O_3$ has been added. Then stop the run.

4.5. Set the temperature of the thermostat about 10℃ higher than that of the first run. Repeat the procedure 4.4.

4.6. The concentration of H_2O_2 should be calibrated by $KMnO_4$ titration before the experiment. Pipette 25ml of H_2O_2 solution into a 150ml conical flask, add 10ml of H_2SO_4, and titrate with $KMnO_4$ till the solution turns light red. Calculate the concentration of H_2O_2 according to the volume of $KMnO_4$ consumed.

5. Data recording and processing

5.1. Fill table 2-15-1 with the data measured and calculated.

5.2. Assuming 25ml of H_2O_2 corresponds to Xml of $Na_2S_2O_3$ standard solution, then we have

$$c_t=\frac{(X-V_{Na_2S_2O_3})\cdot\frac{1}{2}c_{Na_2S_2O_3}}{V+V_{Na_2S_2O_3}} \tag{2-15-8}$$

$$c_0=\frac{(X-1)\cdot\frac{1}{2}c_{Na_2S_2O_3}}{V+1} \tag{2-15-9}$$

$$\frac{c_0}{c_t}=\frac{(X-1)}{V+1}\cdot\frac{V+V_{Na_2S_2O_3}}{X-V_{Na_2S_2O_3}} \tag{2-15-10}$$

where $V_{Na_2S_2O_3}$ is the total volume of $Na_2S_2O_3$ added at time t, $V+V_{Na_2S_2O_3}$ is the total volume of the tested solution at time t, $c_{Na_2S_2O_3}$ is the molar mass concentration of $Na_2S_2O_3$.

In this experiment, $V=500+25=525$ml. c_0 is not necessary the concentration of H_2O_2 at the time when 1ml of $Na_2S_2O_3$ being consumed. The c_0 can be taken after 2ml, 3ml...ml of $Na_2S_2O_3$ have been added and the c_0 in equation (2-15-10) should be 2ml, 3ml...ml accordingly.

5.3. Plot $\ln(c_0/c_t)$ versus t, calculate the rate constants k_1 and k_2 at two temperatures from the slope. Calculate activation energy E_a using equation (2-15-4).

6. Questions

6.1. Under what conditions can the reaction in this experiment be treated as a first order reaction?

6.2. Why solution of $Na_2S_2O_3$ should be added immediately upon the appearance of bluish solution.

6.3. What would be the effect of the amount of KI to be added on the reaction rate? How to explain it?

7. Pharmaceutical Applications

The determination of rate constant and activation energy of chemical reaction is widely used in studies of stabilization and action mechanism of drugs. For example, by measuring the rate constant and activation energy, the mechanism for light induced degradation of ferulic acid was studied to understand the reason why it is unstable in solution, which helps to improve the role of ferulic acid in anti-tumor, anti-hepatic fibrosis and the treatment of cardiovascular and cerebrovascular diseases.

实验十六 丙酮溴化反应速率常数的测定

（一）实验目的

1. 采用初始速率法测定丙酮溴化反应的级数。
2. 室温下测定酸催化下的丙酮溴化反应的反应速率常数。
3. 掌握分光光度计的使用方法。
4. 了解反应速率常数测定的药学应用。

（二）实验原理

在酸溶液中丙酮溴化反应是一个复杂反应。随着溴的消耗，溶液的颜色逐渐由黄变淡，其反应式为

$$CH_3-\overset{\overset{\displaystyle O}{\|}}{C}-CH_3 + Br_2 \rightleftharpoons CH_3-\overset{\overset{\displaystyle O}{\|}}{C}-CH_2Br + Br^- + H^+$$

实验结果表明，在酸度不是很高的情况下，丙酮卤化的反应速率与卤素浓度无关，其速率方程为

$$r=-\frac{dc_{Br_2}}{dt}=\frac{dc_E}{dt}=kc_A^p c_{H^+}^q \qquad 式(2\text{-}16\text{-}1)$$

式中 c_{Br_2} 为溴浓度；c_E 为溴代丙酮浓度；c_A 为丙酮浓度；c_{H^+} 为 H^+ 浓度；k 为反应速率常数；p 和 q 分别为 c_A 与 c_{H^+} 浓度指数。如果 p、q、k 确定，则速率方程也确定。

为测定指数 p,必须进行两次实验。在两次实验中,丙酮的初始浓度不同,而 H^+ 的初始浓度不变。设第一次实验中丙酮的初始浓度 $(c_{A,0})_{Ⅰ}$ 是第二次实验中浓度 $(c_{A,0})_{Ⅱ}$ 的 n 倍,则有

$$\frac{r_{Ⅰ}}{r_{Ⅱ}}=\frac{(-dc_{Br_2}/dt)_{Ⅰ}}{(-dc_{Br_2}/dt)_{Ⅱ}}=\frac{(kc_{A,0}^{p}\cdot c_{H^+}^{q})_{Ⅰ}}{(kc_{A,0}^{p}\cdot c_{H^+}^{q})_{Ⅱ}}=n^{p} \qquad 式(2\text{-}16\text{-}2)$$

将上式取对数,得

$$p=\frac{\lg(r_{Ⅰ}/r_{Ⅱ})}{\lg n} \qquad 式(2\text{-}16\text{-}3)$$

若再做一次实验,使 $(c_{A,0})_{Ⅲ}=(c_{A,0})_{Ⅰ}$,而 $(c_{H^+,0})_{Ⅲ}$ 是 $(c_{H^+,0})_{Ⅰ}$ 的 m 倍,同理可求出指数 q。即

$$q=\frac{\lg(r_{Ⅰ}/r_{Ⅲ})}{\lg m} \qquad 式(2\text{-}16\text{-}4)$$

在实验中,保持丙酮和氢离子的初始浓度远大于溴的初始浓度,那么随着反应的进行,丙酮和氢离子的浓度将基本保持不变。则式(2-16-1)积分后可得

$$-c_{Br_2}=kc_{A}^{p}c_{H^+}^{q}t+Q \qquad 式(2\text{-}16\text{-}5)$$

式中 Q 为积分常数。

本实验由分光光度法在 450nm 处测定溴浓度随时间的变化来跟踪反应的进行。由朗伯-比尔定律,可得

$$A=Bc_{Br_2} \qquad 式(2\text{-}16\text{-}6)$$

式中 A 为吸光度;B 为常数。将式(2-16-6)代入式(2-16-5),得

$$-A=kBc_{A}^{p}c_{H^+}^{q}t+BQ \qquad 式(2\text{-}16\text{-}7)$$

用 A 对 t 作图可得直线,由斜率能求出反应速度常数 k。

(三)仪器和试剂

1. 仪器 分光光度计,超级恒温水浴,100ml 碘量瓶,50ml 容量瓶,5ml、10ml、25ml 移液管,秒表。

2. 试剂 0.02mol/L 溴溶液,4mol/L 丙酮,1mol/L 盐酸。

(四)实验步骤

1. 熟悉分光光度计的构造,学习使用办法,了解注意事项。

2. 仪器校正。

3. 常数 B 的确定 按表 2-16-1 中所列数据,用移液管准确移取 Br_2 溶液和 HCl 溶液于 50ml 容量瓶中,配制 3 份溶液,充分混合放置 10 分钟后,在 450nm 处测它们的吸光度。

4. 丙酮溴化反应动力学参数的确定 将超级恒温水浴温度调至 25℃。按表 2-16-2 设计实验来测定丙酮溴化反应的速率常数和反应级数。精确移取适量的溴、水和盐酸至 100ml 碘量瓶中,混匀。将丙酮溶液精确移入另一碘量瓶中。将两个碘量瓶一起置于 25℃ 恒温水浴中恒温。10 分钟后,将丙酮迅速倒入盛有溴水和盐酸的碘量瓶中,立即开始计时,同时充分混合溶液,并取出一小部分于比色皿中,每分钟在 450nm 处测定一次吸光度,直到吸光度约为 0.1 以下为止。

(五)数据记录和处理

1. 将实验测得值和计算所得 c_{Br_2} 和常数 B 填入表 2-16-1。

表 2-16-1 样品的吸光度和 B 值

Table 2-16-1 Constant B and the Absorbance of Samples

No.	V_{Br_2}/ml	V_{HCl}/ml	V_{H_2O}/ml	c_{Br_2}	A	B	$\bar{B}$
1	10.0	10.0	30.0				
2	6. 0	10.0	34.0				
3	3. 0	10.0	37.0				

2. 按表 2-16-2 设计反应体系。列表记录测得的吸光度,并计算每个吸光度对应的溴浓度。以溴浓度对时间 t 作图,求出同一时刻的反应速度,按式(2-16-3)和式(2-16-4)计算反应级数 p 和 q。

表 2-16-2 初始反应体积

Table 2-16-2 Initial Reactant Volume

No.	V_{Br_2}/ml	V_{HCl}/ml	V_{H_2O}/ml	$V_{丙酮(acetone)}$/ml
1	10	10	20	10
2	10	10	25	5
3	10	5	25	10

3. 以吸光度 A 对时间 t 作图,利用直线斜率,求反应的速度常数 k。

(六)思考题

1. 影响反应速率的主要因素是什么?

2. 本实验中,当反应物丙酮加到含有溴水的盐酸溶液中开始计时,这对实验结果有无影响?为什么?

(七)药学应用

速率常数在药物开发、评价及应用中有重要的应用。在药物动力学研究中,用速率常数来描述药物体内过程速率和浓度的关系,其中吸收速率常数是药物吸收速率与体内药量之间的比例常数,用来衡量药物吸收速度的快慢;消除速率常数则是药物在体内代谢、排泄的速率与体内药量之间的比例常数,用来衡量药物从体内消除速率的快慢。在药物分子设计时,药物分子与靶蛋白的结合速率常数、解离速率常数描述了药物分子和靶蛋白形成的复合物分子之间的不同相互作用,可以为药物分子设计提供更多有用的信息。速率常数可以实验测定,也可通过分子动力学模拟,构建数学模型,与实验方法相辅相成。

Experiment 16 Determination of Rate Constant for the Bromination of Acetone

1. Objective

1.1. To determine the reaction order of bromination of acetone by the method of initial rates.

1.2. To determine the rate constant of acid catalyzed bromination of acetone at room temperature.

1.3. To master the techniques to operate a spectrophotometer.

1.4. To learn the pharmaceutical application of rate constant determination.

2. Principle

In acidic solution, the bromination of acetone is a complex reaction. As the bromine (Br_2) is consumed, the yellow color of the solution fades gradually. The overall stoichiometric reaction equation is as follows

$$CH_3-\overset{\overset{\displaystyle O}{\|}}{C}-CH_3 + Br_2 \rightleftharpoons CH_3-\overset{\overset{\displaystyle O}{\|}}{C}-CH_2Br + Br^- + H^+$$

Experimental results indicate that the reaction rate is independent of the concentration of bromine except at very high acidities. Thus, the rate equation has the form

$$r=-\frac{dc_{Br_2}}{dt}=\frac{dc_E}{dt}=kc_A^p c_{H^+}^q \qquad (2\text{-}16\text{-}1)$$

where k is the rate constant, c_{Br_2} and c_E are concentrations of bromine and bromine-acetone respectively, c_A and c_{H^+} are concentrations of acetone and H^+ ions respectively, p and q are the orders of the reaction with respect to acetone and H^+ ions respectively. Once the p, q and k have been measured, the rate law of the reaction is determined.

In this experiment, the p, q and k are measured by method of initial rates. For measurement of p, two experiments should be performed. In both experiments, the initial concentrations of acetone are changed, while the initial concentrations of H^+ ions hold constant. Assuming the initial concentration of acetone in experiment Ⅰ is n times that of in experiment Ⅱ, say $n=\frac{(c_{A,0})_{\mathrm{I}}}{(c_{A,0})_{\mathrm{II}}}$, we have

$$\frac{r_{\mathrm{I}}}{r_{\mathrm{II}}}=\frac{(-dc_{Br_2}/dt)_{\mathrm{I}}}{(-dc_{Br_2}/dt)_{\mathrm{II}}}=\frac{(kc_{A,0}^p\cdot c_{H^+}^q)_{\mathrm{I}}}{(kc_{A,0}^p\cdot c_{H^+}^q)_{\mathrm{II}}}=n^p \qquad (2\text{-}16\text{-}2)$$

Take logarithm on both sides of the equation (2-16-2), we have

$$p=\frac{\lg(r_{\mathrm{I}}/r_{\mathrm{II}})}{\lg n} \qquad (2\text{-}16\text{-}3)$$

If experiment Ⅲ is run, the reaction order q can also be determined by assuming the initial concentration of H^+ in experiment Ⅰ is m times that of in experiment Ⅲ, say $m=\frac{(c_{H^+,0})_{\mathrm{I}}}{(c_{H^+,0})_{\mathrm{III}}}$, while the initial concentrations of acetone in both runs are the same. Then we have

$$q=\frac{\lg(r_{\mathrm{I}}/r_{\mathrm{III}})}{\lg m} \qquad (2\text{-}16\text{-}4)$$

If the initial concentrations of acetone and H^+ ions are kept much higher than that of bromine, both of the concentrations of acetone and H^+ ion would be considered constant with the progress of the reaction. Integration of equation (2-16-1) gives

$$-c_{Br_2}=kc_A^p c_{H^+}^q t+Q \qquad (2\text{-}16\text{-}5)$$

where Q is the integration constant.

The progress of the reaction can be traced by directly observing the decrease of bromine concentration with spectrophotometric methodology at a wavelength of 450nm. According to the

Lamber-Beer's law, we have

$$A = Bc_{Br_2} \tag{2-16-6}$$

where A is the absorbance, and B is the proportion coefficient. Substitute equation (2-16-5) from equation (2-16-6), it gives

$$-A = kBc_A^p c_{H^+}^q t + BQ \tag{2-16-7}$$

A plot of the absorbance A against t should yield a straight line and the rate constant k can be calculated from the slope of the line.

3. Apparatus and chemicals

3.1. Apparatus

Spectrophotometer, thermostat, 100ml iodine flask, 50ml measuring flask, 5ml, 10ml and 25ml pipettes, and stopwatch.

3.2. Chemicals

0.02mol/L bromine solution, 4mol/L acetone, 1mol/L hydrochloric acid.

4. Procedures

4.1. Be familiar with the construction of spectrophotometer, read the operation manual carefully before use.

4.2. Calibrate the spectrophotometer with distilled water.

4.3. Measurement of constant B

Pipette bromine, water and HCl solution into 3 measuring flasks according to table 2-16-1. Then the mixture is allowed to stand for 10min after mixed thoroughly. Fill the cuvette with above solutions. Determine the absorbance at 450nm.

4.4. Determination of reaction order

Adjust the temperature of the thermostat at 25℃. Use the data listed in table 2-16-2 to run the experiment. Pipette HCl, bromine and water into a 100ml iodine flask, mix thoroughly. Pipette acetone into another 100ml iodine flask. Place both the flask into the thermostat for 10min. Pour the acetone into the mixed solution of HCl and bromine solution, and record the time at once. Mix the solution completely and then quickly transfer a portion of the mixture to the cuvette. Read the absorbance at 450nm every minute until the absorbance is below 0.1.

5. Data recording and processing

5.1. Fill the table 2-16-1 with the absorbance measured to calculate the constant B.

5.2. Design the reaction system according to table 2-16-2. List the absorbance in a table and calculate the concentration of bromine corresponds to each absorbance. Plot the concentration of bromine versus t, and the reaction rate can be calculated to determine the order of the bromine-acetone reaction according to equation (2-16-3) and (2-16-4).

5.3. Plot the absorbance A versus t and calculate the rate constant k from the slope.

6. Questions

6.1. What are the main factors to influence the rate of a chemical reaction?

6.2. In this experiment, timing is started after the addition of acetone into the acidic bromine solution. Does it affect the experiment result? If yes, then why?

7. Pharmaceutical Applications

Rate constant plays an important role in drug developement, evaluation and application. In pharmacokinetics, rate constant characterizes the change of drug concentration in a particular region with time. For example, absorption rate constant is the proportionality constant that relates the rate of drug absorbed into the body to the drug amount at the site of absorption, which is used to describe the rate for the absorption process. While the elimination rate constant is a value used to describe the rate at which a drug is removed from the human system. In drug design, the binding rate constant or the dissociation rate constant of drug with its targe protein are used to predict the correlation of a drug molecule with the protein-ligand complex to provide new ideas for rational drug design. Rate constant can either be determined experimentally or mathematically through molecular modeling to complement each other.

实验十七 最大气泡压力法测定液体表面张力

（一）实验目的

1. 明确溶液表面吸附的概念和特点，理解表面张力和吸附之间的关系。
2. 学会利用最大气泡压力法测定溶液表面张力的原理和技术方法。
3. 根据吉布斯吸附等温式计算溶液表面吸附量，绘制吸附等温线。
4. 了解表面张力的药学应用。

（二）实验原理

1. 表面张力等温式 一定温度下，液体表面张力与溶液浓度之间的关系曲线称为表面张力等温线，如图 2-17-1 所示。若用数学方程式表示表面张力与溶液浓度之间的关系，则称作表面张力等温式。用吸附平衡法可导出表面张力等温式，即

$$\sigma=\sigma_0\left(1-\frac{ac}{1+bc}\right) \qquad 式(2\text{-}17\text{-}1)$$

式中 σ、σ_0 分别为溶液和纯溶剂的表面张力；c 为溶液浓度；a、b 为常数。该式对小分子醇类、羧酸类、酚类（图 2-17-1 中的Ⅱ型曲线）溶液有很好的拟合度。对式(2-17-1)进行线性转换，则可以得到

$$\frac{\sigma_0 c}{\sigma_0-\sigma}=\frac{b}{a}c+\frac{1}{a} \qquad 式(2\text{-}17\text{-}2)$$

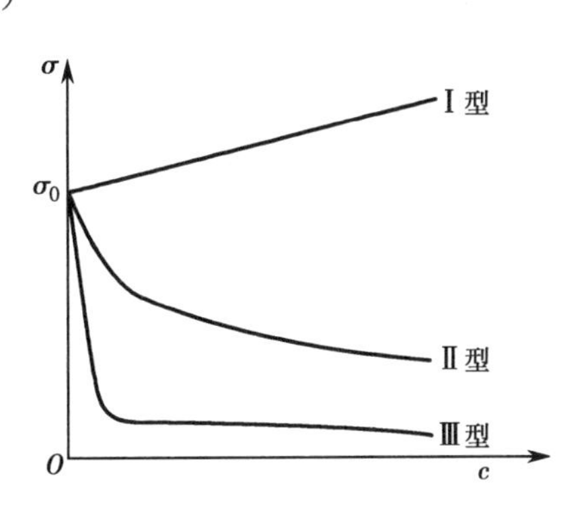

图 2-17-1 表面张力等温线

Fig. 2-17-1 The Isotherm of Surface Tension

2. 吉布斯吸附等温式 溶质在溶液中的分散是不均匀的，也就是说溶质在液体表面层中的浓度和液体内部不同，这种现象称作吸附现象。对于两组分（非电解质）稀溶液，在指定温度与压力下，溶质的吸附量与溶液浓度的关系曲线称表面吸附等温线（图 2-17-2），两者的数学关系服从吉布斯吸附等温式，即

$$\Gamma=-\frac{c}{RT}\left(\frac{\mathrm{d}\sigma}{\mathrm{d}c}\right)_T \qquad 式(2\text{-}17\text{-}3)$$

式中 Γ 为溶质在单位面积表面层中的吸附量；c 为溶液的浓度；T 为热力学温度；σ 为溶液的

表面张力；$\frac{d\sigma}{dc}$称作表面活度。若$\frac{d\sigma}{dc}<0$，溶液的表面张力将随溶质浓度的升高而降低，因此$\Gamma>0$，溶质在液体表面产生正吸附。这类能降低水的表面张力的物质称为表面活性物质。反之，若$\frac{d\sigma}{dc}>0$，溶液的表面张力将随溶质浓度的升高而升高，因此$\Gamma<0$，溶质在液体表面产生负吸附。这类能升高水的表面张力的物质称为表面惰性物质。

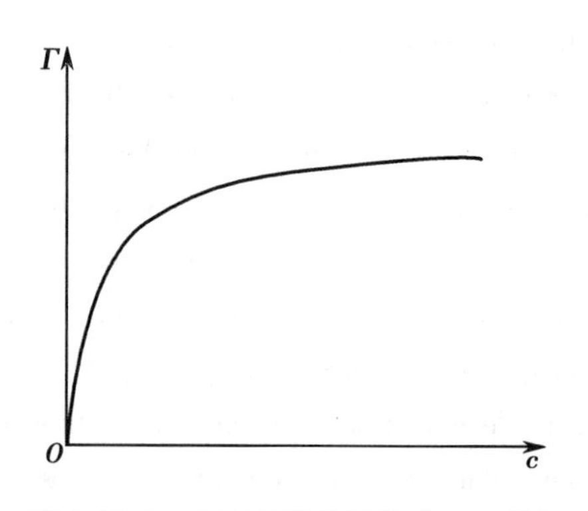

图 2-17-2 表面吸附等温线（正吸附）

Fig. 2-17-2 The Isotherm of Adsorption

对于式(2-17-1)，其相应的吸附等温式为

$$\Gamma=-\frac{c}{RT}\cdot\frac{d\sigma}{dc}=-\frac{c}{RT}\cdot\left(-\frac{\sigma_0 a}{(1+bc)^2}\right)$$

$$=\frac{\sigma_0 a}{RT}\cdot\frac{c}{(1+bc)^2} \qquad 式(2\text{-}17\text{-}4)$$

3. 最大气泡压力法测定表面张力原理　测定表面张力的仪器装置如图 2-17-3a（使用恒温隔套）或图 2-17-3b（使用恒温水浴）所示。

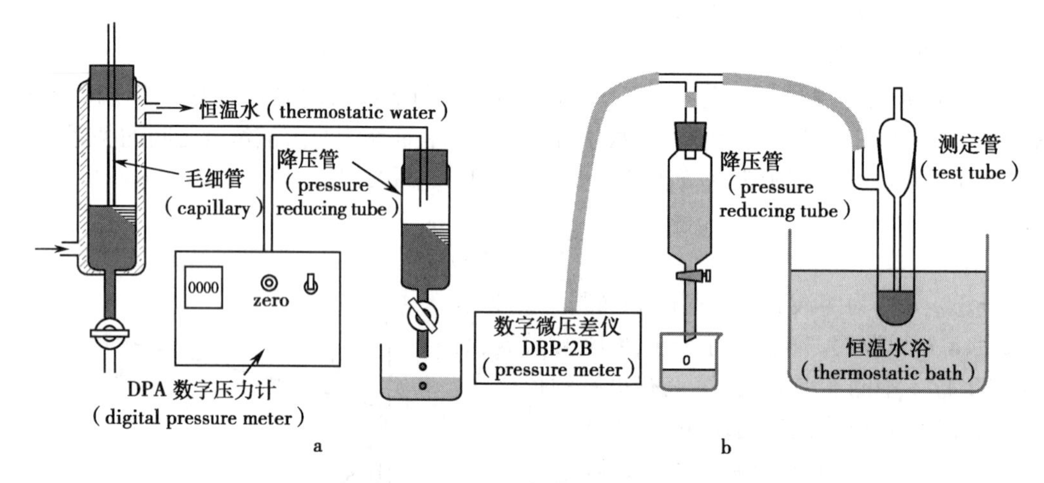

图 2-17-3 最大气泡压力法测定液体表面张力装置示意图

Fig. 2-17-3 Illustration of Apparatus for Surface Tension Measurement by Bubble Pressure Method

测定管中的毛细管端面与液体相切，系统与外压隔开。打开减压装置，使毛细管内溶液受到的压力$p_{外}$大于样品管中液面上的压力$p_{内}$，在毛细管管端缓慢地逸出气泡，毛细管口形成凹液面，产生附加压力$p_r(=p_{外}-p_{内})$。将拉普拉斯公式用于球形液滴，可得

$$p_r=\frac{2\sigma}{r} \qquad 式(2\text{-}17\text{-}5)$$

随着气泡的增大，液面的曲率半径r逐渐减小，p_r逐渐增大。当半球形气泡形成时，r等于毛细管半径R。当气泡继续增大，r又逐渐增大，直至气泡失去平衡而从管口逸出。详细描述见图 2-17-4 所示。

由式(2-17-5)可知，当$r=R$时，p_r有最大值。因此可通过测定气泡逃逸时的最大压力差

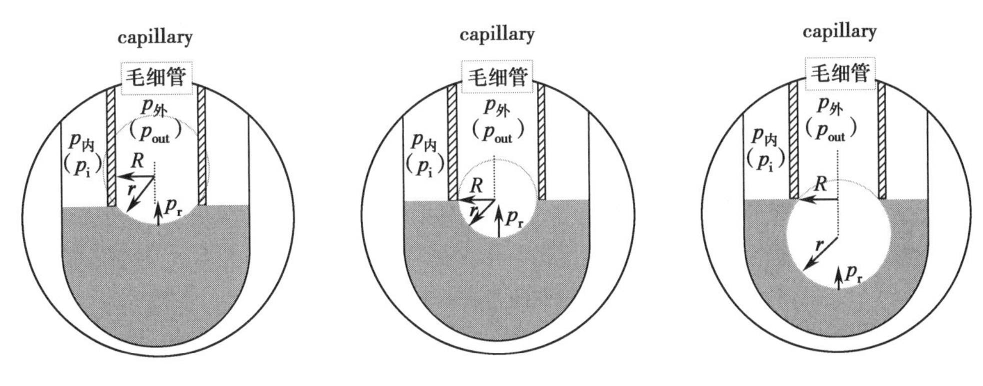

图 2-17-4 毛细管内液面变化和曲面压力变化

Fig. 2-17-4 Variations of Liquid Surface and Variation of Curvature Pressure in Capillary

p_r 来计算表面张力。p_r 可直接由压力计或压差仪读取。

实验温度下水的表面张力 $\sigma_水$ 可由附录表 4-2 得到，用同一毛细管分别测定最大压力 $p_水$ 和 $p_{样品}$，可按下式计算样品的表面张力 $\sigma_{样品}$，即

$$p_{样品}=\frac{2\sigma_{样品}}{R} \quad 式(2\text{-}17\text{-}6)$$

$$p_水=\frac{2\sigma_水}{R} \quad 式(2\text{-}17\text{-}7)$$

$$\sigma_{样品}=\frac{\sigma_水}{p_水}\times p_{样品} \quad 式(2\text{-}17\text{-}8)$$

（三）仪器和试剂

1. 仪器　表面张力测定器（测定管、减压管、精度为 1Pa 的微压差仪），恒温水浴，温度计（量程 0~100℃）。

2. 试剂　醋酸，异戊醇或其他表面活性物质，标准 NaOH 溶液。

（四）实验步骤

1. 配制醋酸水溶液，使其浓度约为 0.1mol/L、0.25mol/L、0.50mol/L、1.0mol/L、1.5mol/L、2.0mol/L 和 3.0mol/L，用标准 NaOH 溶液滴定，确定其准确浓度。

若测定异戊醇溶液的表面张力，配制的溶液浓度分别约为 0.006 25mol/L、0.012 5mol/L、0.025mol/L、0.05mol/L 及 0.1mol/L。用阿贝折光仪测定溶液的折射率，由标准曲线确定溶液的准确浓度。

2. 取一定量的蒸馏水注入事先洗净的测定管中，插入毛细管，调节蒸馏水的量，确保毛细管底端与液面恰好接触。将测定管固定到恒温水浴中，注意保持垂直。调节恒温水浴温度至指定值。压力计调零后，与系统相连。

3. 恒温 5~10 分钟后，打开降压管活塞缓慢放水，系统逐渐减压，控制水的流速使压差计的示值每 1Pa 变化都能显示（约 1 分钟出 8~12 个气泡）。记录气泡逃逸时的最大压差值，连续读取三次（误差不超过±2Pa），取平均值。注意系统不要漏气，液体不能进入连接软管。

4. 按由稀到浓的顺序，同法测定不同浓度的醋酸（或异戊醇）溶液的最大压差值。每次更换溶液时，必须用待测液洗涤毛细管内壁及管壁 3 次，测定管保持相同位置和垂直度。

5. 实验完毕，用洗衣粉（内含去污粉）和热水清洗测定管，蒸馏水冲淋后沥干待用。仪器复位，整理实验台。

6. 注意事项

（1）测定用的毛细管一定要洗干净，否则气泡可能不能连续稳定地逸出，而使压差计读数不稳定，如发生此种现象，毛细管应重洗。

（2）毛细管一定要保持垂直，管口底端刚好插到与液面接触，并且需要控制气泡逸出速度。

（五）数据记录和处理

1. 按表 2-17-1 记录实验数据。

表 2-17-1 最大气泡压力法测定表面张力实验值

Table 2-17-1 Experimental Data of Determination of Surface Tension by Bubble Pressure Method

T:________℃；p:________ kPa

c/（mol/L）	p_r/kPa	σ/（N/m）	Γ/（$\times10^6$ mol/m^3）

2. 以 σ 对 c 作图，得到表面张力等温线（横坐标浓度从零开始）。

3. 用 Excel 软件进行数据拟合。

应用“规划求解”功能对式（2-17-1）进行非线性拟合，求出常数 a 和 b。

也可按方程（2-17-2）作线性拟合，由截距和斜率求出常数 a 和 b。

4. 按式（2-17-4）计算不同溶液浓度的吸附量 Γ，并以吸附量 Γ 对浓度 c 作图，得到表观吸附等温线（在曲线上选择 6~7 个点，对醋酸溶液，用低浓度部分 c<1.5mol/L 进行计算）。

（六）思考题

1. 表面张力测定管的清洁与否对所测数据有何影响？

2. 本实验成败的关键是什么？如果气泡出得很快，或两三个一起出来对结果有什么影响？

3. 毛细管的底端口为什么要刚好接触液面？操作过程中如将毛细管底端口插入液面过深，对结果有何影响？

（七）药学应用

小肠黏膜是药物口服吸收的主要屏障，而小肠黏膜之外的黏液层和水化层是药物到达小肠黏膜的必经之路。例如，研究表明，微乳可以提高难溶性药物的溶解度，较低的表面张力更有利于其透过水化层和黏液层而到达小肠上皮，利于微乳乳滴与小肠上皮细胞的相互作用，增强其跨膜转运的效能，促进难溶药物的口服吸收。再如，外用软膏剂和化妆品是否易于涂敷于皮肤或黏膜而更好地铺展开，主要取决于其基质的表面张力是否低于皮肤或黏膜的临界表面张力。基于此可优化处方。

（八）其他测量系统

研究系统可以改为正丁醇-水或乙醇-水。

正丁醇水溶液，浓度分别为 0.02mol/L、0.05mol/L、0.10mol/L、0.20mol/L、0.25mol/L、0.30mol/L、0.35mol/L 和 0.5mol/L。

乙醇水溶液，浓度分别为 5%、10%、15%、20%、25%、30%、35%、40%。

用正丁醇或乙醇作为研究系统时，需要先用阿贝折光仪测定各配制浓度溶液的折射率，然后根据实验室给出的浓度-折射率标准曲线求得准确的溶液浓度。

Experiment 17 Determination of the Solution Surface Tension by Bubble Pressure Method

1. Objective

1.1. To understand the concept and characteristics of surface adsorption of the solution and the relationship between surface tension and adsorption.

1.2. To learn the principle and technique of the determination of surface tension by the bubble pressure method.

1.3. To calculate the adsorption capacity of the solution surface using Gibbs adsorption isotherm equation and plot the adsorption isotherm curve.

1.4. To learn the pharmaceutical application of surface tension.

2. Principle

2.1. Surface tension isotherm equation

At a specified temperature, the curve which expresses the relationship between surface tension and concentration of the solution is called surface tension isotherm (Fig. 2-17-1). The math equation describing the isotherm is called surface tension isotherm equation. Equation (2-17-1) is one of the isotherm equations derived from adsorption equilibrium method.

$$\sigma=\sigma_0\left(1-\frac{ac}{1+bc}\right) \tag{2-17-1}$$

where σ and σ_0 are surface tensions of the solution and the solvent, respectively, c is the concentration of the solution, a and b are the constants. Equation (2-17-1) is well fitted for small molecular alcohols, carboxylic acids and phenols. The equation can be transformed to a linearity format as follows

$$\frac{\sigma_0 c}{\sigma_0-\sigma}=\frac{b}{a}c+\frac{1}{a} \tag{2-17-2}$$

2.2. Gibbs adsorption isotherm equation

It is known that the surface concentration of solute is different from that of in the bulk, and this phenomenon is called surface adsorption of the solution. For a two-component system, at specified temperature and pressure, the adsorption quantity of the solute is related to the surface tension and the concentration of the solution (Fig. 2-17-2), and can be expressed by Gibbs adsorption isotherm equation

$$\Gamma=-\frac{c}{RT}\left(\frac{\mathrm{d}\sigma}{\mathrm{d}c}\right)_T \tag{2-17-3}$$

where Γ is the surface adsorption quantity of the solute, c is the concentration of the dilute solution, T is the thermodynamic temperature, σ is the surface tension of the solution, and $\frac{\mathrm{d}\sigma}{\mathrm{d}c}$ is the surface activity. When $\left(\frac{\mathrm{d}\sigma}{\mathrm{d}c}\right)_T<0$, then $\Gamma>0$, which is called the positive adsorption, and the substance is a surface active substance. When $\left(\frac{\mathrm{d}\sigma}{\mathrm{d}c}\right)_T>0$, then $\Gamma<0$, which is called the negative adsorption, and the substance is a non-surface active substance.

As to the equation (2-17-1), the corresponding adsorption isotherm equation is

$$\Gamma=-\frac{c}{RT}\cdot\frac{d\sigma}{dc}=-\frac{c}{RT}\cdot\left(-\frac{\sigma_0 a}{(1+bc)^2}\right)=\frac{\sigma_0 a}{RT}\cdot\frac{c}{(1+bc)^2} \quad (2\text{-}17\text{-}4)$$

2.3. The bubble pressure method

In this experiment, the bubble pressure method is used to determine the surface tension of the solution using either of the apparatus shown in Fig. 2-17-3.

Adjust the surface of the liquid to just contact with the bottom of the capillary tubing. Slowly reduce the pressure p_A in the test tube, the pressure p_0 in the capillary tubing will be larger than p_A, leading to the formation of a growing bubble and concave meniscus at the capillary orifice. For a spherical bubble, the additional pressure p_r can be calculated according to Laplace equation. That is

$$p_r=\frac{2\sigma}{r} \quad (2\text{-}17\text{-}5)$$

where σ is the surface tension of the solution, r is the curvature radius of the bubble, and $p_r=p_0-p_A$. With the increase of the bubble, p_r increases with the decrease of radius r of the bubble. When r is exactly corresponding to the radius R of the capillary, the bubble can endure the maximum pressure difference $p_{r,max}$. After the maximum pressure difference, p_A decreases and the radius of the bubble increases until the bubble is detached from the orifice. Fig. 2-17-4 shows the variations of liquid surface and curvature pressure in capillary.

By determining the $p_{r,max}$ at which $r=R$, the surface tension of the solution can be calculated using equation (2-17-5). $p_{r,max}$ can be read on screen of the pressure meter.

At experiment temperature, σ_{water} can be found from Appendice Table 4-2. When the $p_{max,water}$ and $p_{max,sample}$ are determined with the same capillary, the surface tension of the sample can be calculated using the following equation

$$p_{max,sample}=\frac{2\sigma_{sample}}{R} \quad (2\text{-}17\text{-}6)$$

$$p_{max,water}=\frac{2\sigma_{water}}{R} \quad (2\text{-}17\text{-}7)$$

$$\sigma_{sample}=\frac{\sigma_{water}}{p_{max,water}}\times p_{max,sample} \quad (2\text{-}17\text{-}8)$$

3. Apparatus and Chemicals

3.1. Apparatus

Surface tension measuring apparatus include test tube, reduce pressure tube, and micro-pressure meter calibrated to 1Pa, thermostat, thermometer ranged 0~100℃.

3.2. Chemicals

Acetic acid (AR), isoamyl alcohol (AR) or other surfactants (AR), NaOH standard solution.

4. Procedures

4.1. Preparation of acetic acid solution

Prepare a series of acetic acid solutions with the concentrations of approximately 0.1mol/L, 0.25mol/L, 0.50mol/L, 1.0mol/L, 1.5mol/L, 2.0mol/L and 3.0mol/L. Titrate the solutions with standard NaOH to determine the precise concentrations.

When isoamyl alcohol is used, prepare the solutions with the concentrations of approximately 0.006 25mol/L, 0.012 5mol/L, 0.025mol/L, 0.05mol/L and 0.1mol/L, respectively. Determine the refractive index for the solutions by the Abbe refractometer to obtain the precise concentrations of the solutions from a concentration-refractive index working curve.

4.2. Add distilled water into the test tube and insert the capillary tubing tightly into the test tube. Regulate the amount of distilled water in the test tube to make sure that the bottom of the capillary tubing just in contact with the liquid surface. The test tube and the capillary tubing must be kept vertical. Put the test tube in thermostatic water bath of 30℃. Calibrate the pressure meter, and then connect it to the system.

4.3. After constant temperature 5~10 minutes, a thermal equilibrium is reached, slowly open the stopcock of the funnel to discharge water from the bottle. The formation of bubbles must be stable. A suitable rate is about 8~12 bubbles per minute to ensure that each 1Pa difference can be read. Read out the maximum pressure difference on the micro-pressure meter for three times, and then average them. Take care not to allow an air leakage of the system and the entering of liquid into the connection tube.

4.4. According to procedure 4.2 and 4.3, measure the surface tensions of the acetic acid solutions (or isoamyl alcohol) from low to high concentrations. Wash the test tube, especially the capillary tubing with a small portion of the tested solution for 3 times before each measurement. Notice that the test tube must be kept at the same position and vertical through the experiment.

4.5. When the experiment is finished, wash the capillary tubing and the test tube with detergent and hot water, rinse the tube with distilled water and then dry it for use.

4.6. Attentions

4.6.1. Capillary must be clean, otherwise the bubble may not be continuous and steady which cause the unstable readings on pressure meter. Wash the capillary again if such a situation really arises.

4.6.2. Keep the capillary vertical, make sure the nozzle of the capillary is just in contact with the liquid surface, and control the bubble rate.

5. Data recording and processing

5.1. Fill Table 2-17-1 with the data measured and calculated.

5.2. Plot σ versus c with the x-axis starting from 0.

5.3. Use "planning to discriminate" function in Excel to fit equation (2-17-1) and to calculate constant a and constant b. Or according to equation (2-17-2), plot $\frac{\sigma_0 c}{\sigma_0 - \sigma}$ versus c to get a and b from the slope and intercept.

5.4. Calculate the surface adsorption quantity Γ according to equation (2-17-4), and then plot Γ versus c (select 6~7 points from the curve under the concentration of 1.5mol/L when acetic acid solution is tested in the experiment).

6. Questions

6.1. What would be the effect on the result if there are some impurities in the test tube?

6.2. What are the important points concerning a good result in the experiment? What would be the effect on the result when the bubbles run off too fast or two or three bubbles run off together?

6.3. Why should the bottom of the capillary tubing just contact with the liquid surface? Can the capillary be inserted in the interior of the solution for measurement?

7. Pharmaceutical applications

Small intestine mucosa is the main barrier for the oral absorption of drugs, and the mucus layer and hydration layer near the small intestine mucosa is the only way for the drug to reach the small intestine mucosa. For example, it is well known that microemulsion can improve the water solubility of water insoluble drugs, lower surface tension is more conducive to its access to the small intestine epithelial through the hydration layer and mucus layer, to facilitate the interaction between the droplets of the microemulsion and small intestine epithelial cells, thereby enhancing the efficiency of its transmembrane transport, and promoting oral absorption of water insoluble drugs. Furthermore, whether external ointment preparations and cosmetics are easily applied to the skin or mucous membranes and better spread out mainly depends on the surface tension of the matrix. Based on this, the formulation can be optimized.

8. Other model solutions

When *n*-butanol water solution is used, prepare the solutions with the concentrations of 0.02mol/L, 0.05mol/L, 0.10mol/L, 0.20mol/L, 0.25mol/L, 0.30mol/L, 0.35mol/L and 0.5mol/L.

When ethanol water solution is used, prepare the solutions with the concentrations of 5%, 10%, 15%, 20%, 25%, 30%, 35% and 40%.

If either *n*-butanol water solutions or ethanol water solutions are used, measure the solutions by the Abbe refractometer, and find the precise concentrations of the solutions from a concentration-refractive index working curve.

实验十八 固体在溶液中的吸附

（一）实验目的

1. 了解固体在溶液中对溶质吸附量测定的实验方法。
2. 验证弗仑因德立希和兰格缪尔吸附等温式对固体吸附的适用性。
3. 用兰格缪尔吸附等温式求算固体的比表面。
4. 了解固体吸附的药学应用。

（二）实验原理

1. 固体在溶液中的吸附　由于比表面较大的吸附剂（如活性炭、硅胶等）在溶液中有较强的吸附能力，故吸附情况往往比较复杂。这些吸附剂不仅吸附溶质，还吸附溶剂，而且固体、溶质和溶剂三者之间相互作用。因此，在理论上的处理不如气相中对气体的吸附那样简单，吸附量要用表观吸附量$(x/m)_{表观}$表示，即

$$\left(\frac{x}{m}\right)_{表观}=\frac{V(c_0-c)}{m} \qquad 式(2\text{-}18\text{-}1)$$

式中 m 为吸附剂的质量；V 为溶液的体积；c_0 为溶液的起始浓度；c 为吸附达平衡后溶液的浓

度；x 为吸附溶质的摩尔数，该摩尔数并非溶质覆盖在固体表面的绝对值，而是与体相比较所得摩尔数的差值，可正或负，也可能为零。

2. 吸附等温式　对于固体在溶液中的吸附，没有成熟的理论推导公式，一般套用固体在气相中对气体的两个吸附等温式，分别为

$$\left(\frac{x}{m}\right)_{表观}=Kc^{n} \qquad 式(2\text{-}18\text{-}2)$$

$$\left(\frac{x}{m}\right)_{表观}=\frac{\Gamma_{m}bc}{1+bc} \qquad 式(2\text{-}18\text{-}3)$$

式(2-18-2)和式(2-18-3)分别为弗仑因德立希吸附等温式和兰格缪尔吸附等温式，其相应的线性关系式为

$$\ln\left(\frac{x}{m}\right)_{表观}=\ln K+n\ln c \qquad 式(2\text{-}18\text{-}4)$$

$$\frac{c}{(x/m)_{表观}}=\frac{1}{\Gamma_{m}}\cdot c+\frac{1}{\Gamma_{m}b} \qquad 式(2\text{-}18\text{-}5)$$

式(2-18-2)～式(2-18-5)中的 K、n、b、Γ_{m} 均为常数，可通过线性回归方程的截距和斜率求得；Γ_{m} 为饱和吸附量，其物理意义是单位质量或体积的固体的表面所吸附的单分子层的吸附质的量。

由 Γ_{m} 可计算固体的比表面 a，计算式为

$$a=\Gamma_{m}LA \qquad 式(2\text{-}18\text{-}6)$$

式中 L 为阿伏伽德罗常数；A 为溶质分子的截面积，对于直链脂肪酸（包括醋酸），$A=24.3\times10^{-20}\ m^{2}$。按上式测得的比表面，一般比实际值小一些，因为固体表面有部分溶剂的吸附。

（三）仪器和试剂

1. 仪器　100ml 酸式滴定管，250ml 碘量瓶，100ml、150ml 锥形瓶，玻璃漏斗，50ml 碱式滴定管，10ml 和 20ml 移液管，精度为 0.001g 的天平。

2. 试剂　浓度约为 0.04mol/L、0.08mol/L、0.12mol/L、0.20mol/L、0.30mol/L 和 0.40mol/L 的 HAc 溶液（事先标定好），浓度约为 0.1mol/L 的 NaOH 标准液，酚酞指示剂，活性炭。

（四）实验步骤

1. 分别称取 6 份 1.000g 的活性炭于 6 个已编号的碘量瓶中。

2. 分别用 100ml 滴定管在碘量瓶中依次加入 6 个浓度的醋酸溶液 100ml。

3. 振摇碘量瓶约 30 分钟，确保吸附平衡。

4. 按浓度由稀到浓的顺序，分别过滤 6 个碘量瓶中的溶液于锥形瓶中。注意，过滤时要弃去最初部分滤液。

5. 对于浓度较小的三个样品，用移液管吸取 20ml 滤液于 100ml 锥形瓶中，加入酚酞指示剂，用 0.1mol/L 的标准 NaOH 溶液滴定至终点。将消耗的 NaOH 溶液体积除以 2，折算成 10ml 样品的消耗量。每个样品至少滴定 2 次，若 2 次滴定消耗的体积差值大于 0.02ml，应再次滴定。

6. 对于浓度较大的三个样品，用移液管吸取 10ml 滤液，进行同样的滴定分析。

7. 取 10ml 蒸馏水，作空白滴定。

8. 清洗仪器，整理实验台。

（五）数据记录和处理

1. 按表 2-18-1 记录实验数据。

表 2-18-1 醋酸溶液浓度和活性炭对醋酸的吸附量

Table 2-18-1 Concentration of Acetic Acid Solution and Adsorbed Amount of Activated Carbon

m:1.000g;c_{NaOH}:________ mol/L;T:________℃;$V_{样品(sample)}$:10 ml;V_{HAc}:0.1L;V_0:________ ml

No.	c_0/(mol/L)	V_{NaOH}/ml	c/(mol/L)	x/m/(mol/kg)	ln$c$	ln(x/m)	c/(x/m)/(kg/L)
1							
…							
6							

2. 以下式计算平衡浓度 c,将算得到的 c 代入式(2-18-1)计算不同浓度下的表观吸附量,并作活性炭对醋酸溶液的吸附等温线$\frac{x}{m}\sim c$。

$$c=\frac{c_{NaOH}(V-V_0)}{10} \quad \text{式(2-18-7)}$$

3. 以 $\ln\frac{x}{m}\sim\ln c$ 作图和线性回归,求弗仑因德立希吸附等温式中的经验常数 K 和 n。

4. 以$\frac{c}{x/m}\sim c$ 作图和线性回归,求兰格缪尔吸附等温式中的常数 b 和 Γ_m。

5. 由 Γ_m 估算活性炭的比表面。

(六)思考题

1. 实验中为什么过滤活性炭时要弃去部分最初滤液?
2. 实验操作中应注意哪些问题以减少误差?

(七)药学应用

固体在溶液中的吸附在药学研究中有广泛的应用。由于活性炭具有物理和化学性质稳定、耐酸碱、耐高温高压、不溶于水和有机溶剂等诸多优点,可选择吸附液相中的各种物质,在制药行业中常用活性炭除热原、除杂、脱色、助滤等。例如,用活性炭除热原、脱色可改善注射液的澄明度。再如,以杀菌剂碘作为模型药物对交联淀粉微球的吸附性能进行研究。通过改变碘溶液的浓度,调节微球的溶胀能力以及吸附时间来改变交联淀粉微球吸附碘的含量,可获得对碘具有良好吸附性能的交联淀粉微球。

Experiment 18 Adsorption of Solids in Solution

1. Objective

1.1. To learn the method for the determination of the adsorbed amount of solutes by solids in solution.

1.2. To verify the applicability of both Freundlich and Langmuir isothermal adsorption equations in solid adsorptions.

1.3. To calculate the specific surface area of solid by Langmuir isothermal adsorption equation.

1.4. To learn the pharmaceutical application of adsorption of solids in solution.

2. Principle

2.1. Adsorption of solid in solution

Solid adsorbents with larger surface areas, such as activated carbon and silica gel, are effective and often exhibit much more complicated adsorption behavior in solution because they can adsorb both the solute and the solvent and cause interactions among the solid, solute and solvent. Therefore, we have to use apparent adsorption quantity to express their adsorption effectiveness. That is

$$\left(\frac{x}{m}\right)_{\text{Apparent}}=\frac{V(c_0-c)}{m} \quad (2\text{-}18\text{-}1)$$

where m is the mass of the adsorbent, V is the volume of the solution, c_0 is the initial concentration of solution, c is the concentration of solution at adsorption equilibrium, and x is the number of moles of solute adsorbed. The x here is not the absolute moles of solute adsorbed on the surface of the solid, but the mole difference of solute on solid surface and the bulk solution. x could be positive or negative, or even zero.

2.2. Adsorption isotherm

Although there is no well established equation to describe the adsorption of solid in solution, it is found that the adsorption behavior of the solid satisfies the two isothermal equations which are usually applied in adsorption of gases by solids. They are

$$\left(\frac{x}{m}\right)_{\text{Apparent}}=Kc^n \quad (2\text{-}18\text{-}2)$$

$$\left(\frac{x}{m}\right)_{\text{Apparent}}=\frac{\Gamma_{\text{m}}bc}{1+bc} \quad (2\text{-}18\text{-}3)$$

Equations (2-18-2) and (2-18-3) are the Freundlich adsorption isotherm and Langmuir adsorption isotherm respectively. They can be rearranged correspondingly so as to give the linear equations as follows

$$\ln\left(\frac{x}{m}\right)_{\text{Apparent}}=\ln K+n\ln c \quad (2\text{-}18\text{-}4)$$

$$\frac{c}{(x/m)_{\text{Apparent}}}=\frac{1}{\Gamma_{\text{m}}}\cdot c+\frac{1}{\Gamma_{\text{m}}b} \quad (2\text{-}18\text{-}5)$$

where Γ_{m} is the saturated adsorption quantity when the surface of the solid is covered with a layer of unimolecular adsorbates. K, n, b and Γ_{m} in above equations are constants, and can be calculated from the intercepts and slopes, accordingly.

The specific surface area can be calculated from Γ_{m}

$$a=\Gamma_{\text{m}}LA \quad (2\text{-}18\text{-}6)$$

where L is the Avogadro constant, and A is the sectional area of the solute molecule. For linear fatty acid, including acetic acid, $A=24.3\times10^{-20}\ \text{m}^2$. The calculated A is smaller than the real value because of partially adsorbed solid surface by the solvent.

3. Apparatus and Chemicals

3.1. Apparatus

100ml acid burettes, 250ml iodine flasks, 100ml and 150ml conical flasks, glass funnels, 50ml alkali burettes, 10ml and 20ml pipettes, balance calibrated to 0.001g.

3.2. Chemicals

0.04mol/L, 0.08mol/L, 0.12mol/L, 0.20mol/L, 0.30mol/L and 0.40mol/L acetic acid solutions (calibrated), 0.1mol/L standard NaOH solution, phenolphthalein indicator, activated carbon.

4. Procedures

4.1. Weigh accurately 1.000g of activated carbon respectively into 6 numbered iodine flasks.

4.2. Add accurately 100ml of 0.04mol/L, 0.08mol/L, 0.12mol/L, 0.20mol/L, 0.30mol/L and 0.40mol/L acetic acid into above numbered flasks respectively using 100ml acid burettes.

4.3. Shake the flasks for 30min to establish adsorption equilibrium.

4.4. Respectively filter the solutions with glass funnels into conical flasks from diluted to concentrated solutions. Do not forget to discard the initial filtrate.

4.5. Respectively pipette 20ml of the filtrates of three dilute samples into 100ml conical flasks and add with phenolphthalein indicator. Titrate the solution with 0.1mol/L standard NaOH solution. Repeat each titration at least once. Divide the consumed volume of NaOH solution by two to obtain the volume of NaOH solution consumed for titration of 10ml samples. If the volume difference between two titrations is larger than 0.02ml, repeat the titration.

4.6. Respectively pipette 10ml of the filtrates of three concentrated samples into 100ml conical flasks and add with phenolphthalein indicator. Titrate the solution as described above.

4.7. Pipette 10ml of distilled water and repeat procedure 4.5 for a blank titration.

4.8. Clean up the instrument and tidy up the test bench.

5. Data recording and processing

5.1. Fill table 2-18-1 with the data measured and calculated.

5.2. Use the following equation to calculate the concentration at adsorption equilibrium. Substitute the concentration into equation (2-18-1) to calculate the apparent adsorbed amount. Plot the $(x/m)_{\text{apparent}} \sim c$ isotherm for the adsorption of activated carbon to acetic acid.

$$c=\frac{c_{\text{NaOH}}(V-V_0)}{10} \qquad (2\text{-}18\text{-}7)$$

5.3. Plot $\ln(x/m)$ versus $\ln c$ to obtain n and K in Freundlich adsorption isotherm from the slope and intercept of the line.

5.4. Plot $c/(x/m)$ versus c to obtain Γ_m and b in Langmuir adsorption isotherm from the slope and intercept.

5.5. Substitute Γ_m into equation (2-18-6) to calculate the specific surface area of activated carbon.

6. Questions

6.1. Why should the initial filtrate be thrown away when filter the activated carbon in the experiment?

6.2. What precautions should be taken in order to reduce experimental errors?

7. Pharmaceutical applications

Adsorption of solids in solutions is widely used in pharmaceutical research. The activated carbon has many advantages, such as good physical and chemical stability, resistance to acid and alkali, high temperature and high pressure, poor solubility in water and organic solvents, thereby selectively adsorbing various substances in the solutions. The adsorption of activated carbon in solutions is widely used in the pharmaceutical industry such as depyrogenation, impurity removing, decoloration, filtration aiding, and so on. For example, the depyrogenation and decoloration with activated carbon can improve the clarity of the injection.

Another example is to use bactericide iodine as model drug to evaluate the adsorption performance of cross-linked starch microspheres. The better performance on the adsorption of microspheres could be obtained by adjusting concentration of iodine, microsphere's swelling ability, and adsorption time.

实验十九 溶胶的制备及性质

（一）实验目的

1. 学习用不同的方法制备溶胶，并观察溶胶的丁铎尔现象和电泳现象。
2. 学会用界面移动法测定 $Fe(OH)_3$ 溶胶的 ζ 电势和确定电荷符号的方法。
3. 研究电解质对溶胶稳定性的影响。
4. 了解溶胶制备及性质的药学应用。

（二）实验原理

1. 溶胶的制备 通常，溶胶的制备方法有两种，即分散法和凝聚法。分散法是把大颗粒的物质用适当的方法粉碎为胶体大小的质点而获得溶胶；凝聚法是把小分子或离子聚集成胶体大小的质点而制得溶胶。例如，$Fe(OH)_3$ 溶胶就是采用凝聚法制备的。通过水解 $FeCl_3$ 溶液生成难溶于水的 $Fe(OH)_3$，在适当的条件下，过饱和的 $Fe(OH)_3$ 溶液析出小的颗粒而形成 $Fe(OH)_3$ 溶胶。

一般制备所得的溶胶都含有过多的电解质，会使溶胶不稳定，通常可用半透膜渗析、电渗析和超过滤法进行纯化，以除去过多的电解质。

2. 溶胶的光学性质 当把一束可见光投射到分散系统上时，如果分散系统的粒径大于入射光的波长，粒子对光主要起反射作用，如果分散系统的粒径小于入射光的波长，粒子对光主要起散射作用。粗分散系统对可见光主要起反射作用，胶体分散系统对可见光主要起散射作用。当一束可见光通过胶体时，在光线的垂直方向观察，可以看到胶体中有一明亮的光柱，这就是丁铎尔现象。

3. 溶胶的电学性质 通常在相界面上都有双电层存在，根据扩散双电层理论，由于表面分子电离或表面吸附离子而带电荷，在表面附近的介质中必定分布着与表面电性相反而电荷数相等的离子，因而表面和介质间就形成了一定电势差。介质中的离子由于热运动使它们扩散分布着，其中一部分离子紧紧地被吸附在表面形成一定厚度的紧密层，而其余的离子则以扩散的形式分布在表面附近的介质中，形成扩散层。当固液两相发生相对移动时，扩散层离子随介质移动，而紧密层离子则随固相移动，滑动面与介质内部电中性处间的电势差称为 ζ 电势，ζ 电势也叫电动电势。在外加电场的作用下，带电的胶粒与分散的介质间会发生相对运动，胶粒向正极或负极（视胶粒所荷负电或正电而言）移动的现象称为电泳。

测定 ζ 电势对了解溶胶的稳定性具有重要意义，ζ 电势的绝对值越小，则溶胶的稳定性就越差，当ζ电势等于零时，溶胶的稳定性最差，此时可观察到聚沉现象。因此，无论制备溶胶和破坏溶胶，都需要了解所研究溶胶的ζ电势。

原则上，溶胶的任何一种电动现象（电渗、电泳、流动电势和沉降电势）都可以用来测定ζ电势，但常用的还是宏观电泳法。

宏观电泳法的装置如图 2-19-1 所示。ζ电势的计算公式为

$$\zeta=\frac{K\eta u}{4\varepsilon_0\varepsilon_r E} \qquad 式(2\text{-}19\text{-}1)$$

式中K为常数，对于棒状粒子$K=4$，对于球状粒子$K=6$，$Fe(OH)_3$胶粒按棒状粒子处理；ε_0为真空绝对介电常数，其数值为8.854×10^{-12} F/m，ε_r为分散介质的相对介电常数，水的相对介电常数为 81；η为分散介质的黏度，25℃时水的黏度为$0.890\ 4\times10^{-3}$Pa·s；u是电泳速度；E是电势梯度。

电泳速度u可由溶胶界面在外电场作用下迁移的距离h和所用的时间t求出，即$u=h/t$；电势梯度E可由两电极间的电压V和电极间距离l求得，即$E=V/l$。将u、E的表示式代入式(2-19-1)中，并取$K=4$，得

$$\zeta=\frac{\eta hl}{\varepsilon_0\varepsilon_r tV} \qquad 式(2\text{-}19\text{-}2)$$

因此，通过测定溶胶界面在外电场作用下的移动距离，可以计算得到溶胶的ζ电势。

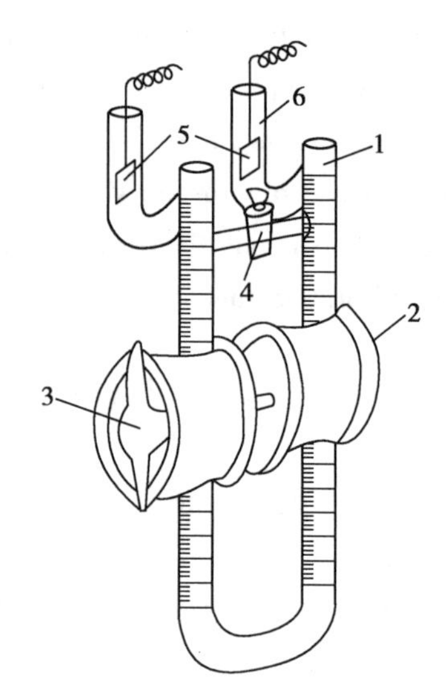

1. U 形管（U tube）；2、3、4. 旋塞（stopcock）；5. 电极（electrode）；6. 弯管（siphon）

图 2-19-1 电泳管示意图
Fig. 2-19-1 Illustration of Electrophoresis Tube

4. 溶胶的稳定性和聚沉作用 溶胶稳定的主要原因是由于胶粒表面带有电荷形成扩散双电层结构，当加入电解质，与溶胶粒子电荷相反的离子压迫扩散层，促使ζ电势趋近于零，粒子可相互接近产生凝聚作用，颗粒逐渐增大而聚沉。

聚沉能力通常用聚沉值表示，聚沉值是使溶胶发生聚沉时需要电解质的最小浓度，其单位用 mol/L。聚沉值与溶胶电荷相反的离子价数 6 次方成反比，即

$$M^{+}:M^{2+}:M^{3+}=\left(\frac{1}{1}\right)^6:\left(\frac{1}{2}\right)^6:\left(\frac{1}{3}\right)^6=100:1.6:0.14 \qquad 式(2\text{-}19\text{-}3)$$

式(2-19-3)称舒尔茨-哈代规则。由此可知，电解质中与溶胶电荷相反的离子价数越高，它的聚沉能力就越强。

（三）仪器和试剂

1. 仪器 电泳管，电泳仪，秒表，25ml 和 100ml 量筒，50ml、200ml 和 1 000ml 烧杯，250ml 三角烧瓶，500W 电炉，铂电极，温度计（100℃），电导率仪，试管，移液管。

2. 试剂 2% 松香乙醇溶液，3% $AlCl_3$ 溶液，0.4mol/L NH_4OH，0.1mol/L HCl，3% $FeCl_3$ 溶液，5% 火棉胶液，1% $AgNO_3$ 溶液，1mol/L KCl 溶液，0.025mol/L K_2SO_4 溶液，0.015mol/L $K_3[Fe(CN)_6]$溶液。

（四）实验步骤

1. 渗析袋的制备 在一洁净、干燥的 250ml 锥形瓶中，倒入约 20m1 5% 火棉胶液（注意远离火焰），小心转动锥形瓶，使其在瓶壁形成一均匀薄层。将多余的火棉胶液倒回瓶中（注

意勿倒入水池),把锥形瓶倒置在铁圈上,让剩余的火棉胶液流尽。待溶剂挥发完,轻触火棉胶膜不再黏手时,在锥形瓶内加满自来水溶去残留的溶剂,几分钟后倒掉水,在锥形瓶口剥离开一部分膜。在膜与瓶壁之间注入自来水,一个完整的渗析袋就脱离瓶壁。取出袋后,将其加满蒸馏水,检查是否漏水,如果袋完好无损,泡入蒸馏水中备用。

2. 溶胶的制备

(1) 置换溶剂法制备松香水溶胶:在一 50ml 烧杯中加入 20ml 蒸馏水,在搅拌下逐滴加入 5~6 滴 2% 松香乙醇溶液(若松香过多会形成浑浊的悬浮液)。观察松香水溶胶的丁铎尔现象、透视光和侧射光,并与松香乙醇溶液对比。

(2) 胶溶法制备 $Al(OH)_3$ 溶胶:在一 50ml 烧杯中加入 5ml 3% 的 $AlCl_3$ 溶液,加入略微过量的 0.4mol/L NH_4OH 溶液。将 $Al(OH)_3$ 沉淀过滤出,并用蒸馏水洗涤数次。当沉淀成极黏稠时,将其转入 100ml 蒸馏水中煮沸,并不时滴加 3~4 滴 0.1mol/L HCl。继续加热约 10 分钟后,停止加热,放置约 1 小时,沉淀几乎全部胶溶,观察丁铎尔现象。

(3) 水解法制备 $Fe(OH)_3$ 溶胶:取 200ml 蒸馏水于烧杯中,加热至沸腾,边搅拌边滴加 20ml 3% 的 $FeCl_3$ 溶液,再煮沸约 2 分钟,得棕红色的 $Fe(OH)_3$ 溶胶。取约 20ml $Fe(OH)_3$ 溶胶放入小烧杯中,用 20ml 蒸馏水稀释后,观察丁铎尔现象。剩余的溶胶留作电泳和聚沉实验。

3. 溶胶的渗析 将 $Fe(OH)_3$ 溶胶倒入火棉胶袋中,扎好袋口,将其放入一盛有约 600ml 蒸馏水的 1 000ml 烧杯中。根据情况选用热渗析法或冷渗析法纯化溶胶。

(1) 热渗析:加热至 60~70℃,30 分钟换一次水,直至渗析液用 $AgNO_3$ 溶液检查不出 Cl^- 时为止。

(2) 冷渗析:在室温下静置 3~4 天,中间更换一两次蒸馏水。

4. 溶胶的聚沉作用 用移液管在三个干净的试管中各注入 2ml $Fe(OH)_3$ 溶胶,然后在每个试管中分别用滴管慢慢地滴入 1mol/L KCl、0.025mol/L K_2SO_4 和 0.015mol/L $K_3[Fe(CN)_6]$ 溶液,并摇动。注意,在开始有明显聚沉物出现时,即停止加入电解质,记下每次所用的溶液滴数。

5. ζ 电势的测定 将纯化好的 $Fe(OH)_3$ 溶胶倒入一 200ml 的干净烧杯中,测其电导率。另取一干净的 200ml 烧杯,配制稀盐酸。先加入 150ml 蒸馏水,再往里滴加 0.1mol/L 的盐酸,一直加到与溶胶的电导率相同为止。用少许溶胶洗涤电泳管一次,然后将溶胶注入电泳管,使液面高于旋塞 2、3,关闭旋塞 2、3,倒去旋塞上面多余的 $Fe(OH)_3$ 溶胶,用配制的稀盐酸溶液洗涤两次,然后将稀盐酸溶液加满电泳管。将电泳管固定到铁架上。开通旋塞 4,使两侧溶液平衡。调节液面至弯管高度 2/3 处。将两个铂电极插入弯管里。关闭旋塞 4,同时开启旋塞 2、3,使溶胶与稀盐酸溶液之间形成清晰的界面。接通电源,调节输出电压约为 200V,观察溶胶与稀盐酸溶液之间界面的迁移。待界面上升至旋塞上方有刻度部分时,选定某一位置作为界面上升的起始位置开始计时,等界面上升约 2.5cm 时,记下界面上升的距离 h 和所消耗的时间 t 以及加在两电极间的电压 V。关闭电源,倒掉电泳管中的液体,用金属丝量取两电极间的距离(电流在液体中流过的距离)l。实验结束后,洗净电泳管,待下次实验使用。

(五)数据记录与处理

1. 记录各实验现象,并加以解释说明。

2. 试比较三种电解质聚沉值的大小,说明 $Fe(OH)_3$ 溶胶所带电荷。

3. 按表 2-19-1 记录实验数据。

表 2-19-1 电泳实验数据

Table 2-19-1 Data of Electrophoresis Experiment

t/s	h/m	V/V	l/m

4. 将数据代入式(2-19-2),计算 $Fe(OH)_3$ 溶胶的 ζ 电势。

(六) 思考题

1. 影响电泳速度的因素有哪些?
2. 本实验中所用稀 HCl 的电导率为什么必须和所测溶胶的电导率十分相近?
3. 电解质引起溶胶发生聚沉作用的根本原因是什么?

(七) 药学应用

溶胶的制备与性质在药学实践中的应用是十分广泛的,比如混悬剂的制备和稳定性研究、中药活性成分的提取、胶体金检测等。ζ 电势测定和电泳技术在药学实践中也有很大价值。比如在制药行业中最难的问题之一是药物稳定的分散在合适的悬浮载体中,一种制备高稳定的药物胶体分散体系的有效方法是借助电泳迁移率测定将体系设计成具有最大分散度。另外,溶胶可以克服疏水性药物在配制和递送方面的困难,甚至改善跨生物屏障的不良药物的渗透性,有研究将具有药理协同作用的疏水性药物与紫杉醇封装在聚合物胶束中,该策略解决了前者水溶性差,结晶速度快等缺点,经静脉给药后发现具有良好的治疗作用。

Experiment 19 Preparation and Properties of Sols

1. Objective

1.1. To learn the different methods for the preparation of sols, and to observe the Tyndall phenomenon as well as the electrophoresis phenomenon of the sols.

1.2. To learn the measurement of zeta (ζ) potential for $Fe(OH)_3$ sol by interface moving method and to determine the sign of the electric charge of $Fe(OH)_3$ sol.

1.3. To investigate the influence of electrolytes in the stability of sols.

1.4. To learn the pharmaceutical application of preparation and properties of sols.

2. Principle

2.1. Preparation of sols

Sols can be prepared by dispersion or coacervation methods. The dispersion method is that big particles are disintegrated into colloidal dimension. The latter is that small molecules or ions are coacervated into colloidal size. For example, $Fe(OH)_3$ sol can be obtained by the coacervation method. Hydrolysis of $FeCl_3$ in aqueous solution forms sparingly soluble $Fe(OH)_3$ at first. Then under suitable conditions, small particles of $Fe(OH)_3$ are separated out from supersaturated $Fe(OH)_3$ solution to form $Fe(OH)_3$ sol.

The prepared sols usually contain appreciable quantities of electrolytes which will result in instability of the sols. Therefore, it is necessary to remove them from sols. There are generally three methods available for sol purification, *i.e.*, dialysis, electrodialysis, and ultrafiltration.

2.2. The optical properties of sols

The dispersion system of the colloid has the ability to scatter light. When a light beam direct at a colloidal dispersion, its path through the sol may be illuminated by the light that scattered by the colloidal particles. This is known as the Tyndall effect. Tyndall scattering occurs when the dimensions of the particles that are causing the scattering are larger than the wavelength of the radiation that is scattered.

2.3. The electric properties of sols

Generally, there is an electrical double layer at the interface of two phases. According to the theory of diffuse electrical double layer, the surface shall carry electric charges due to the molecular ionization or the ions adsorption occurring at the surface. As a consequence, counter ions with the same quantity of opposite charges in the medium distribute around the surface. Thus potential difference arises between the surface and the medium. As a result of equilibrium between electrostatic attraction and thermal motion, part of ions are closely adsorbed at the surface to form a compact double layer, and the rest distribute in the form of diffusion in the medium to form a diffuse double layer. When a solid phase moves relatively to the liquid phase, ions in the diffuse double layer move with the medium, and ions in the compact double layer move with the solid phase. The potential difference between the sliding surface and the neutral region of the solution is called ζ potential. In an external electric field, a charged colloidal particle would move relative to the dispersed medium. This phenomenon is called electrophoresis.

Measuring ζ potential is of great significance for knowing the stability of a sol. The lower the absolute value of ζ potential is, the less stable is the sol. A ζ potential of zero value corresponds to minimum stability and coagulation of the sol will take place.

In principle, any of the electrokinetic phenomena (electroosmosis, electrophoresis, streaming potential, and sedimentation potential) can be used for measurement of ζ potential. However, the commonly used method is apparent electrophoresis.

The experimental apparatus of the method is illustrated in Fig. 2-19-1. ζ potential can be calculated using the following equation

$$\zeta=\frac{K\eta u}{4\varepsilon_0\varepsilon_r E} \tag{2-19-1}$$

where K is a constant ($K=4$ for rod-like particles, $K=6$ for spherical particles, and $Fe(OH)_3$ sol is regarded as rod-like particles), ε_0 is the vacuum permittivity with the value of 8.854×10^{-12} F/m, ε_r is the relative permittivity of disperse medium ($\varepsilon_{r,water}=81$), η is the viscosity of the disperse medium [$\eta_{water(298K)}=0.890\ 4\times10^{-3}$ Pa · s], u is the electrophoresis rate, and E is the potential gradient.

The electrophoresis rate u can be calculated by the moving distance h of the interface of the sol in the electrophoresis tube and the time consumed t, i.e. $u=h/t$. The potential gradient E can be calculated by the voltage and the distance between two electrodes, i.e. $E=V/l$. Take $K=4$, and substitute h/t and V/l for u and E in equation (2-19-1) respectively, we have

$$\zeta=\frac{\eta hl}{\varepsilon_0\varepsilon_r tV} \tag{2-19-2}$$

Therefore, by measuring the interface movement of the sol, the ζ potential can be calculated.

2.4. The stability and coagulation effect of sols

The major reason that the sols can be stable is the existence of the electrical double layer at the interface. When electrolyte solution is added to the sols, the ions of the electrolyte solution with the opposite electric charge repress the diffuse double layer and ζ potential will be close to zero. Then colloidal particles can approach closely enough for van der Waals forces to hold them together and allow aggregation. The coagulation is generated by the increase of the colloidal particle size.

The capacity of coagulation is usually expressed by coagulation values which is the lowest concentration of electrolyte solution at which the coagulation is generated. The relationship between coagulation value and the valence of the counter-ion follows the Schulze-Hardy rule. The polyprecipitation value is inversely proportional to the ion valence opposite to the sol charge to the power of 6, that is

$$M^+ : M^{2+} : M^{3+}=\left(\frac{1}{1}\right)^6 : \left(\frac{1}{2}\right)^6 : \left(\frac{1}{3}\right)^6=100 : 1.6 : 0.14 \tag{2-19-3}$$

Equation (2-19-3) indicates that higher valence of the counter-ion gives a lower coagulation concentration and reflects the higher capacity of coagulation.

3. Apparatus and chemicals

3.1. Apparatus

Electrophoresis tube, electrophoretic instrument, stopwatch, measuring cylinders with capacities of 25ml and 100ml, respectively, beakers with capacities of 50ml, 200ml, and 1 000ml, respectively, 250ml conical flask, 500W electric cooker, platinum electrode, thermometer (100℃), conductometer, test tube, pipette.

3.2. Chemicals

2% colophony-ethanol solution, 3% $AlCl_3$ solution, 0.4mol/L NH_4OH solution, 0.1mol/L HCl solution, 3% $FeCl_3$ solution, 5% collodion, 1% $AgNO_3$ solution, 1mol/L KCl solution, 0.025mol/L K_2SO_4 solution, 0.015mol/L $K_3[Fe(CN)_6]$ solution.

4. Procedures

4.1. Preparation of dialysis bag

Pour about 20ml of 5% collodion into a 250ml clean and dried conical flask (away from the flame). Tilt the flask while turn it carefully and slowly to form a thin layer of collodion along the inner surface of the conical flask. Spill the rest collodion back into the bottle. Do not pour it into the basin. Invert the conical flask and put it on an iron loop to drop out all the residual collodion and volatilize the solvent. When the solvent has passed off and the collodion membrane has no longer viscidity, pour tap water in the conical flask to dissolve the rest solvent and then pour out. Peel off part of the membrane at the mouth of the conical flask, fill tap water between the membrane and the inner wall of the conical flask to form an integrated dialysis bag, and then take it out. Check the bag to make sure there is no hole in it by pouring water in it, and then dip it into distilled water for use.

4.2. Preparation of sols

4.2.1. Preparation of colophony-water sols by the solvent substitution method: Add about 20ml of distilled water in a 50ml beaker, drop in 5 ~ 6 drops of 2% colophony-ethanol solution with stirring. Observe the Tyndall phenomenon, transmission light, and the sidelight for the sol and compare it with 2% colophony-ethanol solution.

4.2.2. Preparation of sols by the peptization method: Add 5ml of 3% $AlCl_3$ solution in a 50ml beaker, add a little excessive amount of 0.4mol/L NH_4OH to the above solution. Filtrate out the deposition of $Al(OH)_3$, wash it with distilled water several times, transfer it into 100ml of distilled water to heat to boil for 10min, and add 3~4 drops of 0.1mol/L HCl between whiles. Stop heating, and allow to stand for 1h for a complete peptization. Then observe the Tyndall phenomenon.

4.2.3. Preparation of sols by the hydrolytic method: Add about 200ml of distilled water into a beaker, heat it to a boil state, and drop in 20ml of 3% $FeCl_3$ solution with stirring. Boil the solution for about 2min to obtain wine $Fe(OH)_3$ sol. Take out about 20ml of $Fe(OH)_3$ sol to a small beaker, dilute the sol with 20ml of distilled water, and then observe the Tyndall phenomenon. Keep the rest of the $Fe(OH)_3$ sol for electrophoresis and coagulation experiments.

4.3. dialysis of the sols

Pour the $Fe(OH)_3$ sol into the dialysis bag, tie it up, and put it in a 1 000ml beaker containing about 600ml distilled water. Choose the hot or cold dialysis method accordingly to purify the sol.

4.3.1. Hot dialysis: Heat the water to 60~70℃, change the distilled water every half an hour until no chloride ions being detected in the dialysate.

4.3.2. Cold dialysis: Allow the beaker to stand at room temperature for 3~4 days, and change the distilled water once or twice in the period.

4.4. Coagulation effect of sols

Add 2ml of $Fe(OH)_3$ sol in each of three clean test tubes using the pipettes, then slowly drop in 1mol/L KCl, 0.025mol/L K_2SO_4, and 0.015mol/L $K_3[Fe(CN)_6]$ to three test tubes with shaking, respectively. Stop dropping in the electrolyte solution and record the number of solution drops when the obvious coagulation effect begins to appear.

4.5. Measurement of ζ potential

Transfer the purified $Fe(OH)_3$ sol to a 200ml clean beaker, measure the conductivity of the sol. Add about 150ml of distilled water to another beaker, drop in 0.1mol/L HCl with stirring till the conductivity of the solution is the same as that of the $Fe(OH)_3$ sol. Wash the electrophoresis tube first with a small amount of $Fe(OH)_3$ sol once, and then pour rest of the sol into the tube till the sol surface is over the stopcocks 2 and 3. Close the stopcocks, pour away the sol over the stopcocks, wash the tube with a small amount of the dilute HCl twice, and fill the tube with the dilute HCl. Fix the electrophoresis tube, turn on the stopcock 4, adjust the level of the solution to the height of 2/3 of the siphons, and insert electrodes to each of the siphons. Close the stopcock 4, turn on the stopcocks 2 and 3 at the same time to form clear interface between $Fe(OH)_3$ sol and dilute HCl. Turn on the power supply, set the voltage to about 200V. Observe the movement of the interface and choose a reference mark to time. When the interface has moved up about 2.5cm, record time and moved distance of the interface, and voltage. Turn off the power supply, pour away the solution in the tube, and measure the distance between the two electrodes with an iron wire. Clean the electrophoresis tube for later use.

5. Data recording and processing

5.1. Record the phenomenon of the experiment and explain them.

5.2. Compare the coagulation values of three different electrolyte solutions and indicate the electric charge of $Fe(OH)_3$ sol.

5.3. Fill table 2-19-1 with the data measured.

5.4. Calculate the ζ potential for $Fe(OH)_3$ sol with equation(2-19-2).

6. Questions

6.1. What are the factors affecting the electrophoretic speed?

6.2. Why the conductivity of the dilute hydrochloric acid must be much nearer to that of the measured sol?

6.3. What is the basic reason for the coagulation effect of sols due to the electrolyte solutions?

7. Pharmaceutical application

The preparation and properties of sols are used widely in pharmaceutical practice. For example, study on preparation and stability of suspensions, extracting the active ingredient of traditional Chinese medicine, colloidal gold test. ζ potential measurements and electrophoresis are also very valuable in pharmaceutical practice. For example, one of the most difficult problems in the pharmaceutical industry is the preparation of physically stable dispersions of a drug in a suitable suspension vehicle. A highly stable drug colloidal dispersion system can be obtained after the conditions of maximum dispersion are established utilizing electrophoretic mobility measurements. In addition, sol can overcome the difficulties in the preparation and delivery of hydrophobic drugs, and even improve the permeability of drugs that hardly across biological barriers. Some studies have encapsulated hydrophobic drugs with pharmacological synergy with paclitaxel in polymer micelles. This strategy solves the disadvantages of the former, such as poor water solubility and fast crystallization speed. It was found to have good therapeutic effect after intravenous administration.

实验二十　黏度法测定大分子的平均相对分子质量

（一）实验目的

1. 掌握黏度法测定大分子化合物平均相对分子质量的原理和实验方法。

2. 掌握乌贝路德黏度计的特点及测量原理。

3. 了解黏度测定的药学应用。

（二）实验原理

黏度法是目前应用最广泛的测定大分子平均相对分子质量的方法。大分子稀溶液的黏度反映了液体在流动时存在着的内摩擦。常用的黏度表示方法及物理意义分别为：

（1）溶剂黏度 η_0：溶剂分子间的内摩擦表现出来的黏度。

（2）溶液黏度 η：溶剂分子间、大分子间和大分子与溶剂分子间三者内摩擦的综合表现。

（3）相对黏度 η_r：$\eta_r=\dfrac{\eta}{\eta_0}$，溶液黏度与溶剂黏度的比值。

（4）增比黏度 η_{sp}：$\eta_{sp}=\dfrac{\eta-\eta_0}{\eta_0}=\eta_r-1$，溶液黏度比溶剂黏度增加的相对值。

（5）比浓黏度$\dfrac{\eta_{sp}}{c}$：单位浓度下所显示出的黏度。

（6）特性黏度$[\eta]$：$\lim\limits_{c\to 0}\dfrac{\eta_{sp}}{c}=[\eta]$，无限稀溶液中大分子与溶剂分子之间的内摩擦。

大分子的相对分子量愈大，则它与溶剂间的接触表面也愈大，内摩擦力也愈大，表现出的特性黏度也就愈大。特性黏度和大分子的相对分子质量间的经验关系式为

$$[\eta]=KM_\eta^\alpha \qquad 式(2\text{-}20\text{-}1)$$

式中 M_η 为黏均分子质量；K 和 α 是与温度、大分子和溶剂的性质有关的常数。α 还与分子的形状和大小有关，其数值一般介于 0.5~1 之间。K 与 α 的数值可通过渗透压或光散射等方法求得。通过测定大分子溶液的黏度求得特性黏度$[\eta]$，代入式(2-20-1)就可以计算出大分子化合物的平均相对分子质量。

黏度的测定有许多方法，其中以乌贝路德黏度计法最为方便。若测得液体从毛细管黏度计中的流出时间，通过泊稷叶公式可计算黏度。当溶液流出的时间 t 大于 100 秒时，泊稷叶公式可表示为

$$\eta=\frac{\pi pr^4t}{8LV} \qquad 式(2\text{-}20\text{-}2)$$

式中 η 为液体的黏度；r 为毛细管的半径；t 为溶液流出的时间；p 为毛细管两端的压力差（$p=\rho gh$，ρ 为液体的密度，g 为重力加速度，h 为毛细管中液体的平均高度）；L 为毛细管的长度；V 为流经毛细管的液体体积。

设 η_0、ρ_0、t_0 分别为溶剂的黏度、密度和一定体积溶剂流经毛细管的时间，η、ρ、t 分别为待测液体的黏度、密度和同体积待测液流经同一毛细管的时间，则根据式(2-20-2)，溶剂和待测液体的黏度之间的关系为

$$\frac{\eta}{\eta_0}=\frac{pt}{p_0t_0} \qquad 式(2\text{-}20\text{-}3)$$

下标“0”表示溶剂，将 $p=\rho gh$ 代入式(2-20-3)，可以得到

$$\frac{\eta}{\eta_0}=\frac{\rho t}{\rho_0t_0} \qquad 式(2\text{-}20\text{-}4)$$

若测定在稀溶液中进行，$\rho\approx\rho_0$，则式(2-20-4)可改变为

$$\eta_r=\frac{\eta}{\eta_0}=\frac{t}{t_0} \qquad 式(2\text{-}20\text{-}5)$$

当溶液浓度趋近于零时，$[\eta]=\lim\limits_{c\to 0}\dfrac{\eta_{sp}}{c}=\lim\limits_{c\to 0}\dfrac{\ln\eta_r}{c}$。所以，测定 η_r 后，即可算出 η_{sp}/c 和 $\ln\eta_r/c$，再根据下列经验式，即

$$\frac{\eta_{sp}}{c}=[\eta]+\beta_1[\eta]^2c \qquad 式(2\text{-}20\text{-}6)$$

$$\frac{\ln\eta_r}{c}=[\eta]-\beta_2[\eta]^2c \qquad 式(2\text{-}20\text{-}7)$$

用η_{sp}/c和$\ln\eta_r/c$对浓度c作图，线性外推得到[η]。如图2-20-1所示，两条线应重合于一点，由此也可以校验实验的可靠性。

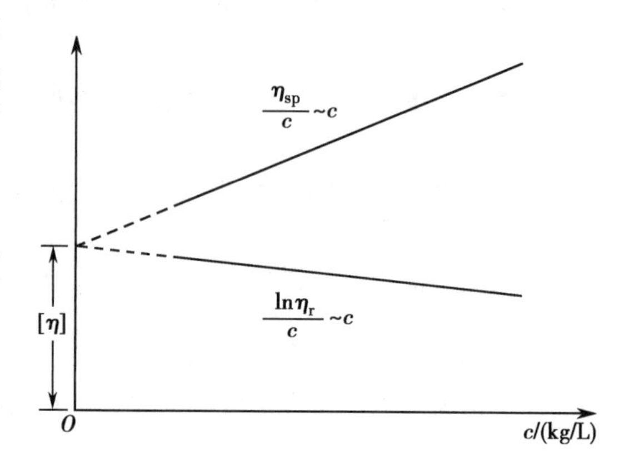

图2-20-1 外推法求[η]

Fig. 2-20-1 The Extrapolation to Get [η]

（三）仪器与药品

1. 仪器 恒温槽，乌贝路德黏度计，秒表，5ml和10ml移液管，100ml注射器，3号砂芯漏斗，50ml容量瓶，50ml烧杯，100ml锥形瓶，250ml吸滤瓶，洗耳球，螺旋夹，橡皮管（长约20cm）。

2. 药品 右旋糖酐（分析纯，平均相对分子量：40 000~60 000）或聚乙烯醇（分析纯）。

（四）实验步骤

1. 右旋糖酐溶液的配制 用分析天平准确称取1.5g右旋糖酐样品，倒入预先洗净的50ml烧杯中，加入约30ml蒸馏水，在水浴中加热溶解至溶液完全透明，取出自然冷却至室温，再将溶液移至50ml的容量瓶中，加蒸馏水至刻度（浓度为0.03kg/L）。然后用预先洗净并烘干的3号砂芯漏斗过滤，装入100ml锥形瓶中备用。

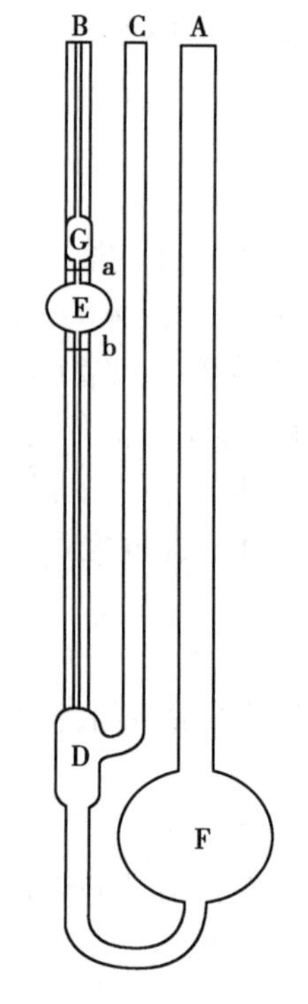

图2-20-2 乌贝路德黏度计

Fig. 2-20-2 Ubbelohde Viscometer

2. 黏度计的清洗 先用洗液将黏度计洗净，再用自来水、蒸馏水分别冲洗几次，尤其注意要反复冲洗毛细管部分，洗好后烘干备用。

3. 仪器的安装 图2-20-2为乌贝路德黏度计示意图。调节恒温槽温度至25℃，按照实验装置，如图2-20-3所示，在黏度计的B管和C管上都接上橡皮管，然后将黏度计垂直放入恒温槽并固定，使水面完全浸没G球。

4. 溶液流出时间的测定 用移液管量取10ml右旋糖酐溶液（c_1），放入干燥洁净的黏度计中，恒温10分钟。用夹子夹紧C管上的橡皮管使之不通气，用注射器经橡皮管从B管抽气，待到液面上升至G球的一半时，移去注射器，立即打开C管上的橡皮管的夹子，D球以上的液体悬空。毛细管中的液体在重力作用下流下，当液面降到刻度a时，开启秒表计时，待液面到达刻度b时，停止计时，此即液体流过a-b刻度所需时间t。重复操作三次，每次相差不大于0.3秒，取三次的平均值为t_{av1}。然后用移液管由A管加入5ml蒸馏水，使溶液浓度为c_2，用注射器将溶液混合均匀，并将溶液通过毛细管和G球部荡洗2~3次，再测定液体流过所

需时间 t_{av2}。同样依次准确加入 5ml、5ml、10ml、10ml 蒸馏水稀释溶液，溶液浓度分别为 c_3、c_4、c_5、c_6，用同样的方法分别测定液体流过 a-b 刻度所需的时间 t_{av3}、t_{av4}、t_{av5}、t_{av6}。

5. 溶剂流出时间的测定 倒掉黏度计中的溶液，用自来水洗净黏度计，尤其要反复冲洗黏度计的毛细管部分。最后用蒸馏水清洗 3 次。在黏度计中加入约 10~15ml 蒸馏水，按照上述方法测定其流出时间 t_{av0}。实验完毕后，倒掉黏度计中的液体，洗净，收拾好仪器和实验台。

6. 实验步骤的其他设计方法 配制 0.06kg/L的右旋糖酐，先取 10ml 蒸馏水测定溶剂的 t_{av0}，然后加入 10ml 右旋糖酐测定 t_{av1}，再依次加 5ml、5ml、5ml、10ml 蒸馏水测定 t_{av2}、t_{av3}、t_{av4}、t_{av5}，其优点是不必取出黏度计，保持测定条件的恒定。

7. 注意事项

（1）黏度计必须洁净干燥。

（2）测量过程中黏度计必须垂直放置，且在实验过程中不得晃动黏度计。

（3）样品在恒温槽中恒温后方可测量。

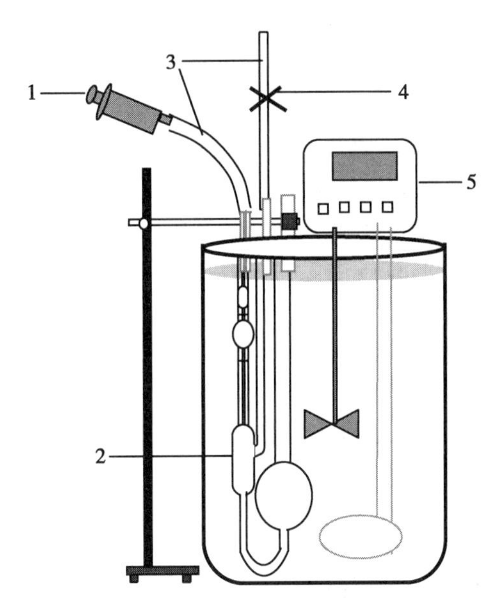

1. 注射器（injector）；2. 乌贝路德黏度计（Ubbelohde viscometer）；3. 橡皮管（rubber tubes）；4. 螺旋夹（screw clamp）；5. 恒温槽（thermostat）

图 2-20-3 黏度测定实验装置图
Fig. 2-20-3 Sketch of Apparatus for Viscosity Measurement

（五）数据记录与处理

1. 实验数据记录（表 2-20-1）

表 2-20-1 不同浓度大分子溶液的黏度和流经毛细管的时间
Table 2-20-1 Viscosities and Traveling Time through Capillary of Macromolecule with Different Concentrations

溶剂（solvent）：________；T：________℃

c/（kg/L）	t_{av}/s	η_r	$\frac{\eta_{sp}}{c}$	$\frac{\ln \eta_r}{c}$
c_0				
c_1				
…				
c_6				

2. 作 $\eta_{sp}/c\sim c$ 及 $\ln\eta_r/c\sim c$ 图，并外推到 $c\to0$，由截距求出$[\eta]$，或作线性回归，求取回归直线的截距和相关系数。一般由 $\eta_{sp}/c\sim c$ 得到的数据较为可靠。

3. 由式(2-20-1)计算右旋糖酐的黏均相对分子质量 M_η。在 25℃时，右旋糖酐水溶液的 $K=9.78\times10^{-2}$L/kg，$\alpha=0.5$。若测定聚乙烯醇的黏均相对分子质量，25℃时，$K=2.0\times10^{-2}$L/kg，$\alpha=0.76$。

（六）思考题

1. 为什么用特性黏度$[\eta]$来求算大分子的平均相对分子质量？它和纯溶剂黏度有区别吗？

2. 乌贝路德黏度计的毛细管太粗或太细对测定结果有何影响？

（七）药学应用

溶液的黏度在药物制剂中具有非常重要的意义。例如，黏度是半固体制剂处方设计及制备工艺过程优化的关键物理参数之一，合适的黏度可以保证良好的药物释放；许多药用辅料如羟丙基甲基纤维素等就采用黏度进行分级。此外，一些眼部药物常通过提高制剂黏度来提高黏膜黏附性，以此延长药物的停留时间，有研究证明，较高的黏度能使具有相似粒径的吲哚美辛悬浮液在眼部的吸收提高数倍，从而显著提高药物的生物利用度。

（八）其他测定方法

1. 转筒法　在两同轴圆筒间充以待测液体，外筒作匀速转动，测内筒受到的黏滞力矩。

2. 落球法　通过测量小球在液体中下落的运动状态来求黏度。

Experiment 20 Determination of Average Relative Molar Mass of the Macromolecule by Viscosity Method

1. Objective

1.1. To understand the principle of the viscosity method in molar mass measurement and to master the experimental technique.

1.2. To understand the characteristics and the measuring principle of the Ubbelohde viscometer.

1.3. To learn the pharmaceutical application of viscosity determination.

2. Principle

The viscosity method is widely used to measure the average relative molar mass of macromolecules due to its convenient procedure, simple requirements of equipment, and accurate results. For the dilute macromolecule solution, the viscosity reflects the internal resistance to flow. The most common solution viscosity terms and their meanings are listed as follows.

(1) Viscosity of the solvent, η_0: It represents the internal frictional force that arises between solvent molecules.

(2) Viscosity of the solution, η: It represents the internal frictional forces that arise between solvent molecules, between solute molecules, and between both the solute and solvent molecules.

(3) Relative viscosity, η_r: $\eta_r=\frac{\eta}{\eta_0}$, it is the ratio of the solution viscosity to solvent viscosity.

(4) Specific viscosity, η_{sp}: $\eta_{sp}=\frac{\eta-\eta_0}{\eta_0}=\eta_r-1$, it is the fractional increase in the viscosity over that of the pure solvent caused by the addition of the macromolecule.

(5) Reduced viscosity, $\frac{\eta_{sp}}{c}$: It is the viscosity at the unit concentration.

(6) Intrinsic viscosity, $[\eta]$: $\lim\limits_{c\to 0}\frac{\eta_{sp}}{c}=[\eta]$. It represents the internal frictional force between

both the solute and solvent molecules, where the solution is at infinite dilution.

The bigger the relative molar mass of the macromolecule is, the larger contacted area between the molecules of the solute and the solvent there is. Thus, the inner friction and the value of the intrinsic viscosity are higher as well. The experiential relation between the intrinsic viscosity and the relative molar mass of the macromolecule, M, is

$$[\eta]=KM_{\eta}^{\alpha} \tag{2-20-1}$$

where M_{η} is the viscosity average molar mass, K and α are constants related to temperature, properties of macromolecule, and solvent. The α also depends on the shape and size of the macromolecule with the value of 0.5 ~ 1.0. The values of K and α can be obtained by ways of osmosis and light scattering, and so on. The intrinsic viscosity can be obtained by measuring the viscosity of the solution and the solvent, and then substituted into equation (2-20-1) to calculate the relative molar mass of the macromolecule.

There are many ways to measure viscous behavior. A convenient measurement is based on Ubbelohde viscometer. When the flowing time of the liquid through the capillary of the viscometer is measured, the viscosity of the solution can be calculated by the Poiseuille equation

$$\eta=\frac{\pi pr^{4}t}{8LV} \tag{2-20-2}$$

where η is the viscosity of the solution, r is the radii of the capillary, t is the time required for the solution to flow through the capillary, p is the pressure difference between the two ends of the capillary, L is the length of the capillary, and V is the volume of the solution flow through the capillary. It is known that $p=\rho gh$, where ρ is the density of the solution, g is the gravitational acceleration, and h is the average height of the liquid column in the capillary.

Whenthe time is taken for the same volumes of the solution and the solvent to flow through the same capillary is measured respectively, according to equation (2-20-2), the relationship between viscosities of solvent and solution is expressed as follows

$$\frac{\eta}{\eta_0}=\frac{pt}{p_0t_0} \tag{2-20-3}$$

The subscript "0" in equation (2-20-3) refers to solvent. Then substitute ρgh for p in equation (2-20-3), we have

$$\frac{\eta}{\eta_0}=\frac{\rho t}{\rho_0t_0} \tag{2-20-4}$$

For a dilute solution, $\rho\approx\rho_0$, then equation (2-20-4) becomes

$$\eta_r=\frac{\eta}{\eta_0}=\frac{t}{t_0} \tag{2-20-5}$$

It can be proved that $[\eta]=\lim\limits_{c\to0}\frac{\eta_{sp}}{c}=\lim\limits_{c\to0}\frac{\ln\eta_r}{c}$ as c tends to zero. After determining η_r, the η_{sp}/c and $\ln(\eta_r/c)$ can be calculated respectively. Then according to the following empirical equations

$$\frac{\eta_{sp}}{c}=[\eta]+\beta_1[\eta]^2c \tag{2-20-6}$$

$$\frac{\ln\eta_r}{c}=[\eta]-\beta_2[\eta]^2c \tag{2-20-7}$$

by plotting η_{sp}/c versus c, $\ln\eta_r/c$ versus c, the $[\eta]$ can be calculated from the intercepts of the two straight lines. Fig 2-20-1 indicates that the two intercepts are equivalent.

3. Apparatus and chemicals

3.1. Apparatus

Thermostat, Ubbelohde viscometer, stopwatch, 5ml and 10ml pipettes, 100ml syringe, 3# core filler funnel, 50ml measuring flask, 50ml beaker, 100ml conical flask, 250ml filter flask, rubber suction ball, screw clamp, rubber tubes (about 20cm in length).

3.2. Chemicals

Dextran (AR, average relative molar mass of 40 000~60 000), or polyvinyl alcohol (AR).

4. Procedures

4.1. Preparation of dextran solution

Weigh out 1.5g of dextran precisely using an analytical balance and put it into a 50ml beaker, add about 30ml of distilled water into the beaker. Heat the beaker in a bath till the dextran is dissolved completely, take out the beaker from the bath and cool it unaffectedly to room temperature. Transfer the solution into a 50ml measuring flask, add distilled water to volume to prepare the solution with the concentration of 0.03kg/L. Filter the solution through a clean dry 3# core filler funnel, and transfer the filtrate into a 100ml measuring flask for use.

4.2. Cleaning of the viscometer

Immerse the viscometer in chromic acid mixture, and rinse it first with tap water, then with distilled water for several times. The capillary must be rinsed carefully and repeatedly. Dry up the viscometer for use.

4.3. Installation of the apparatus

Adjust the temperature of thermostat to 25℃. Connect two rubber tubes to the tubes B and C of the viscometer respectively, put the viscometer into the thermostat to have the G ball of the viscometer soaked under the water and fix it. Fig. 2-20-2 is the illustration of the Ubbelohde viscometer. Fig. 2-20-3 is the sketch of apparatus for viscosity measurement.

4.4. Determination of flow time of the solution through the capillary

Take 10ml of the dextran solution (c_1) into the viscometer using a pipette. After thermal equilibrium for 10min, clamp the rubber tube on tube C. Deflate the air in tube B with a syringe. When the solution level in tube B is raised to about half of the G ball, take off the injector and unclamp the rubber tube immediately. Therefore, the solution over ball G is hanged in the air. The solution in the capillary gravitates downwards. When the solution level in tube B falls to scale "a", start to time. As long as the solution level in tube B falls to scale "b", stop timing. It is the time taken for the solution between scales "a" and "b" to flow through the capillary. Repeat the steps three times and the time difference between each measuring must not exceed 0.3s. Average the three results and take it as t_{av1}. Then add 5ml of distilled water in the viscometer through tube A with a pipette to obtain a solution of concentration c_2. Mix the solution well with the syringe, have the solution passed through the capillary and ball G 2~3 times, and then repeat the steps of t_{av1} measurement to determine t_{av2}. Add 5ml, 5ml, 10ml, and 10ml distilled water to dilute the solution, the concentrations of the solutions are c_3, c_4, c_5, and c_6 respectively. Repeat the steps as t_{av2} measurement to determine t_{av3}, t_{av4}, t_{av5}, and t_{av6} correspondingly.

4.5. Determination of flow time of the solvent through the capillary

Pour away the solution in the viscometer. Rinse the viscometer with tap water first and then with distilled water several times. The capillary region must be rinsed repeatedly and carefully. Add about 10~15ml of distilled water in the cleaned viscometer, use the same method as described in procedure 4.4 to determine the flowing time of the solvent through the capillary, and record it as t_{av0}. Pour away the liquid in the viscometer after the experiment is finished. Clean up the apparatus and desk.

4.6. Other experimental step design

Prepare the dextran solution with a concentration of 0.06kg/L. Pipette 10ml of distilled water into the viscometer and measure the t_{av0} firstly. Then add 10ml of the dextran solution and measure the t_{av1}. After this measuring, add 5ml, 5ml, 5ml, and 10ml of distilled water to the viscometer in turn and measure the t_{av2}, t_{av3}, t_{av4}, and t_{av5} correspondingly. The advantage of this design is that, without taking out the viscometer, the conditions for measurement are still steadily maintained.

4.7. Attentions

4.7.1. The viscometer must be clean and dry.

4.7.2. During the measurement, viscometer must be placed vertically, and should be stay still.

4.7.3. The sample can only be measured until the temperature in thermostatic bath is constant.

5. Data recording and processing

5.1. Fill table 2-20-1 with the data measured and calculated.

5.2. Plot η_{sp}/c versus c and $\ln\eta_r/c$ versus c to calculate $[\eta]$ from the extrapolated intercepts at $c=0$. Or calculate the intercept and correlation coefficient by linear regress. Commonly the result is more credible from $\eta_{sp}/c \sim c$.

5.3. Calculate the average relative molar mass of the dextran by equation (2-20-1). At 25℃, for dextran in water, $K=9.78\times10^{-2}$ L/kg and $\alpha=0.5$. For polyvinyl alcohol in water, $K=2.0\times10^{-2}$ L/kg and $\alpha=0.76$ at 25℃.

6. Questions

6.1. Why use intrinsic viscosity $[\eta]$ to calculate the average relative molar mass of the macromolecule? Are there any differences between the viscosity of the pure solvent and the intrinsic viscosity?

6.2. What would be the effect on the results if the capillary of the Ubbelohde viscometer is too thick or too thin?

7. Pharmaceutical applications

The viscosity of the solution is important in pharmaceutics. For example, viscosity is one of the key physical parameters in formulation design and optimization of half solid formulation preparation. The proper maintenance of viscosity is necessary to ensure good drug delivery; many pharmaceutical excipients, such as hydroxypropyl methyl cellulose, are classified by viscosity. In addition, the mucosal adhesion of some ocular drugs is often improved by improving the viscosity of the preparation, so as to prolong the residence time of the drug. Studies have proved that higher viscosity can increase the absorption of indomethacin suspension with similar particle size several times in the eye, thus significantly improving the bioavailability of the drug.

8. Other measurement methods

8.1. Rotation cylinder method

After the liquid is filled between coaxial-cylinder, the outside cylinder turns with a constant torque; the viscosity torque of the inside cylinder is measured.

8.2. Falling ball method

8.3. Viscosity is measured by the motion state of a small ball in liquid.

实验二十一　药物多晶型的差热分析（设计性实验）

（一）实验目的

1. 培养学生独立思考、综合运用知识设计实验的能力。

2. 熟悉药物多晶型的制备和表征技术。

3. 了解药物多晶型的特性及药学研究的重要性。

（二）实验背景

同一种物质的分子形成多种晶型的药物，称为药物多晶型，亦称同质异晶现象。对于固体制剂、半固体制剂、混悬剂等剂型，应注意药物是否存在多晶型。因为同一固体药物可因结晶条件、制备工艺不同而得到不同晶型的晶体。由于药物晶型的不同，其物理化学性质也不同，直接影响药物的质量与药效，因此在新药申报生产过程中，需要进行实验研究。

药物的多晶型研究可加强新药研发过程中对化合物的生理活性和生物利用度的确认，给药途径、剂型选择及相应工艺的优化，从而全方位提高药物质量，对保证药物的理化性质稳定、生物利用度提高、治疗效果增强和毒性减小都有着重要意义。故对多晶型进行准确、快速的鉴别很有必要。

目前鉴别晶型的方法主要是针对不同晶型具有不同的理化特性及光谱学特征来进行的熔点测定法、红外分光光度法、X 射线衍射法和热分析法等。相对于目前常用的红外光谱分析与 X 射线衍射法，热分析方法不但可以测定药物的多晶型，还可以测得其晶型是单向型晶型还是可逆型晶型。借助于相关公式，可以进行晶型转换的动力学计算，计算活化能和转化速率，因而热分析法已经成为药物晶型研究中常用的方法。药物多晶型的形成因素复杂，溶剂、温度、湿度和压强的改变都可以导致晶型的形成和转变。

（三）实验提示

1. 选用不同的溶剂、溶液浓度、结晶条件与速度等，制备药物的不同晶型。

2. 查阅相关文献，了解多晶型的制备和表征的测定技术。文献查阅时使用的关键词：药物多晶型、多晶型制备、晶型转变、多晶型确定或鉴别。

3. 根据实验室条件，找到合适的方法，制备药物多晶型并对制备的晶体进行晶型鉴别。

（四）实验室提供的条件

1. 仪器　CRY-4P 型差热分析仪，红外光谱仪，X 射线衍射仪，核磁共振仪，电源，计算机，打印机，其他常用玻璃仪器等。

2. 试剂　桂美辛，西咪替丁，α-Al_2O_3（分析纯，200 目），Sn 粉（200 目），丙酮，甲醇，乙醇,苯,乙醚,氯仿及其他常用和必须试剂等。

(五)实验记录和报告

实验报告应包括实验设计的基本原理、实验方法和实验步骤、原始数据记录、实验数据的处理方法和结果、依据实验方案和结果自行设计的讨论题目(应包括对实验原理的分析、实验的影响因素、实验中出现的问题和可能的原因、实验方案改进的设想、实验结论等),以及参考文献。

(六)实验要求

1. 利用各种检索工具查阅相关文献并做出较为详细的摘录。
2. 参考相关文献,通过本人的综合思考,拟订详细的实验方案,并独立实施。
3. 对实验结果进行详细的归纳总结。

Experiment 21 Thermal Analysis of Polymorphism Pharmaceuticals (Designing Experiment)

1. Objective

1.1. To train the ability of scientific consideration and experimental designing.

1.2. To familiar with the preparation and characterization of polymorphism of pharmaceutical components.

1.3. To learn the properties and significance of polymorphism in pharmaceutical research and application.

2. Background

Polymorphism is the ability of a solid material to exist in two or more crystal forms with different arrangement or conformations of the constituents in the crystal lattice. It is potentially to be found in solid preparations, semisolid preparation, suspensions and other dosage forms. Polymorphism can be obtained by different crystallization conditions and the preparation process. The polymorphic forms of a drug differ in the physicochemical properties, which directly affect the quality and efficacy of the drug. Therefore, polymorphic studies are much important, especially in new drug application.

Polymorphism research can enhance the confirmation of physiological activity and bioavailability and optimization of route of administration, dosage form and technology in new drug research, thus improving the drug qualities like stable physical and chemical properties, bioavailability improvement, enhancement of therapeutic effect and toxicity reduction. Therefore, accurate and rapid identification of crystal forms is very necessary.

At present the methods of identifying the crystal form are mainly the melting point method, infrared spectrophotometry, X-ray diffraction and thermal analysis, etc. Compared with infrared spectrum analysis and X-ray diffraction method, the thermal analysis method can not only determine drug polymorphism, but also can distinguish monotropic crystal or enantiotropic crystal. And thermal analysis method has become a commonly used method in drug crystal research.

The formation and transformation of polymorphism is affected by solvent, temperature, humidity, pressure.

3. Some suggestions

3.1. The polymorphic forms of drug can be prepared with different solvent, concentration of solution, crystallization condition and rate, etc.

3.2. Consult references to understand the properties of polymorphism. Learn and borrow the method from studies of polymorphism suggested in 3.1. or other references. Use the following key words to consult references: polymorphism, preparation of polymorphism, transformation of polymorphism, characterization of polymorphism.

3.3. Prepare and characterize the polymorphic crystals according to laboratory condition.

4. Apparatus and chemicals

4.1. Apparatus

CRY-4P type differential thermal analyzer, infrared (IR) spectrometer, X-ray diffractometer, nuclear magnetic resonance (NMR) spectrometer, power supply, computer, printer, other commonly used glass instruments.

4.2. Chemicals

Cinmetacin, Cimetidine, α-Al_2O_3 (AR, 200 mesh), Sn powder (200 mesh), acetone, alcohol, ethanol, benzene, ethyl ether, chloroform etc.

5. Experimental report

Write the report after finishing the experiment. The report should include the principle of the experiment, experimental method and procedures, data record and processing, results and discussion (the evaluation of experimental design, the influence factors, problems arising from experiment and possible reasons, improved vision for the experiment, conclusion etc.), and references.

6. Requirements

6.1. Use any index tools to find the information concerned.

6.2. According to the references and laboratory condition, design the experimental procedures, carry them out and analyze the data independently.

6.3. Discuss the result with high standards.

实验二十二 药物稳定性及有效期测定（设计性实验）

（一）实验目的

1. 培养学生独立思考、独立设计实验的能力。
2. 掌握化学反应动力学方程和温度对化学反应速率常数的影响。
3. 熟悉药物的结构特点及其影响稳定性的因素。
4. 了解药物含量测定方法，设计化学动力学实验。

（二）实验背景

药品的稳定性是指原料药及制剂保持其物理、化学、生物学和微生物学性质的能力。稳定性研究贯穿药品研究与开发的全过程，一般始于药品的临床前研究，在药品临床研究期间和上市后还应继续进行稳定性研究。我国《化学药物稳定性研究技术指导原则》中详细规定了样品的考察项目、考察内容以及考察方法,不仅为药品的生产、包装、贮存、运输条件和有效期的确定提供了科学依据,也保障了药品使用的安全有效性。

在药品稳定性研究中,原药含量是考察项目之一,一般以原药量降低5%(特殊规定除外)的时间定为药物贮存有效期。在实际工作中,通常需要快速有效的方法预测药物制剂的稳定性,从而为进一步研究工作提供基础。

本设计实验要求学生自己查阅相关文献,结合实验室的实际情况,选择合适的实验方法,自主设计实验方案,独立完成实验操作和数据处理,对某一药物的贮存有效期进行预测。

(三)实验提示

1. 药物选择金霉素水溶液(pH=6),或维生素C注射液。

2. 查阅相关文献,了解药物的性质和特点,了解和借鉴他人研究金霉素水溶液或维生素C化学稳定性的方法。文献查阅时使用的关键词(供参考):金霉素、维生素C、稳定性、有效期。

3. 找到合适的分析方法,为动力学研究提供基础。

4. 通过加温加速实验,预测药物稳定性。通过快速实验得到的数据计算正常条件下药物的贮存有效期。

5. 设计实验方案时,应充分注意实验室的条件。

(四)实验室提供的条件

1. 仪器 紫外-可见分光光度计,超级恒温水浴,各种常用玻璃仪器。

2. 药品 金霉素水溶液(pH=6),维生素C注射液,其他常用试剂。

(五)实验报告

实验报告应包括实验设计的基本原理、实验方法和实验步骤、原始数据记录、实验数据的处理方法和结果、依据实验方案和结果自行设计的讨论题目(应包括对实验原理的分析、实验的影响因素、实验中出现的问题和可能的原因、实验方案改进的设想、实验结论等)和参考文献。

(六)实验要求

1. 利用各种检索工具查阅相关文献并做出较为详细的摘录。

2. 参考相关文献,通过本人的综合思考,拟订详细的实验方案,并独立实施。

3. 对实验结果进行详细的归纳总结。

Experiment 22 Determination of Drug Stability and Shelf Life (Designing experiment)

1. Objective

1.1. To train the ability of scientific consideration and experimental designing.

1.2. To master the chemical kinetic equation for simple order reaction and the influence of temperature on reaction rate constant.

1.3. To familiar with relationship between drug structure and its stability.

1.4. To learn the determination method of drug content, and to design chemical kinetic experiment independently.

2. Background

Drug stability refers to the ability of raw material drugs or formulations to keep their properties in physics, chemistry, biology and microbiology. Stability study is of great importance in drug study and development. Although drug stability study is usually developed in the period of preclinical stage, it still needs to be developed in the periods of clinical study and marketing. The observation items, contents and methods, influence factors for stability study are explained in *Governing Principle for the Technological Study of the Stability of Chemical Drug*, which not only providing scientific rationale for drug production, package, storage, transportation and stability determination, but also ensuring safety and effect on drug usage.

In drug stability study, the content of raw material drug is one of the observation items. Usually the drug shelf life is defined as the length of raw material drug contents to reduce to 95% of its initial value. In practice, a quicker and more effective method of drug stability study is preferred.

This designing experiment requires the students to consult references, to choose appropriate experimental method according to laboratory condition, to design the experimental procedures, to carry them out and to analyze the data independently.

3. Some suggestions

3.1. Model drug Aureomycin solution at pH = 6, or vitamin C injection.

3.2. Consult references to understand the properties and characteristics of the drugs. Learn and borrow the method for study of aureomycin or vitamin C from the references. Use the following key words to consult references: aureomycin, vitamin C, stability, shelf life.

3.3. Establish an appropriate analyzing method.

3.4. Predict the stability of the drug studied by accelerated testing, and calculate the shelf life of the drug at normal condition.

3.5. Consider the laboratory condition to design the experiment.

4. Apparatus and chemicals

4.1. Apparatus

Ultraviolet-visible spectrophotometer, super thermostat, commonly used glass instrument.

4.2. Chemicals

Aureomycin solution at pH = 6, or vitamin C injection, other regents.

5. Experimental report

Write the report after finishing the experiment. The report should include the principle of the experiment, experimental method and procedures, data record and processing, results and discussion (the evaluation of experimental design, the influence factors, problems arising from experiment and possible reasons, improved vision for the experiment, conclusion etc.), and references.

6. Requirements

6.1. Use any index tools to find the information concerned.

6.2. According to the references and laboratory condition, design the experimental procedures, carry them out and analyze the data independently.

6.3. Discuss the result with high standards.

实验二十三 载药纳米粒子的制备及表征（综合设计性实验）

（一）实验目的

1. 培养学生独立思考、综合运用所学知识设计实验的能力。
2. 熟悉载药纳米粒子的制备方法及表征测试技术。
3. 了解纳米粒子的特性和应用背景。

（二）实验背景

纳米粒子是指粒径在1~100nm的微粒或固体，具有小尺寸效应、表面效应和宏观量子隧道效应，能产生特有的电学、磁学、光学和化学等特性，在诸多领域中具有重要的应用价值，受到人们的广泛重视。近年来，纳米技术在医药学领域也取得了飞速的发展。药物经直接纳米化或采用纳米载体系统载药所制备的新型药物制剂，无论是在药物的理化性质，还是药代动力学和药效动力学特征方面均与常规制剂有明显的差异。缓释、控释、靶向特性提高了药物的生物利用度与治疗指数，而且新药研究与开发的成功率也有所提高，因而该领域成为药学研究的前沿和热点之一。

用于载药的纳米粒子多种多样，如脂质体、聚合物胶束、纳米粒、纳米囊等，也可以将药物分散成纳米晶体。纳米粒子的制备工艺和方法包括薄膜分散法、溶剂乳化蒸发法、高压均质法、胶体磨法等。纳米粒子应具有较高的药物包封率，这样才能保证药物以纳米的形式发挥生物学作用，此外，载药率也应满足用药要求。

载药纳米粒子的物理化学性质与其生物学性质密切相关。制备得到的载药纳米粒子首先需对其物理化学性质进行评价。一般情况下，主要性质的评价包括载药率、药物包封率、外观形态、粒径分布和ζ电势等。评价粒子的形态和尺寸的手段很多，例如利用扫描电镜或透射电镜观察纳米粒子的形状和大小；利用动态光散射法可以测定纳米粒子的粒径大小及其分布以及ζ电势。

本设计实验要求学生自己查阅相关文献，结合实验室的实际情况选择合适的实验方法、自主设计实验方案、独立完成实验操作和数据处理。

（三）实验提示

1. 纳米粒子的制备　可采用磷脂、胆固醇制备脂质体，采用pH梯度法或$(NH_4)_2SO_4$梯度法装载多柔比星，制备多柔比星脂质体；也可以采用PEG-PLA等制备聚合物胶束，薄膜分散法装载多柔比星；也可以采用单硬脂酸甘油酯、吐温80等制备固体脂质纳米粒装载脂溶性药物等。

2. 查阅相关文献，了解载药纳米粒的制备方法、纳米制剂中药物含量的测定方法以及粒径和形态的测试技术。文献查阅时使用的关键词（供参考）：脂质体、固体脂质纳米粒、聚合物胶束、制备、性质评价。

3. 根据实验室条件，选择合适的制备方法和评价方法，对载药纳米粒子的形貌、粒度和载药率等进行分析。

（四）实验室提供的条件

1. 仪器　旋转蒸发仪，紫外-可见分光光度计，X射线衍射仪，透射电镜，扫描电镜，激光粒度仪，高速离心机，电磁搅拌器，其他常用玻璃仪器等。

2. 药品 磷脂，胆固醇，PEG-PLA，单硬脂酸甘油酯，吐温 80，多柔比星，其他常用和必须试剂。

（五）实验记录和报告

实验记录和报告应包括实验设计的基本思想、实验方法和实验步骤、原始数据记录、实验数据的处理方法和结果、依据实验方案和结果自行设计的讨论题目（应包括对实验设计的评价、影响实验结果的因素、实验中出现的问题和可能的原因、实验方案改进的设想等）和参考文献。

（六）实验要求

1. 利用各种检索工具查阅相关文献并做出较为详细的摘录。
2. 参考相关文献，通过本人的综合思考，拟订详细的实验方案，并独立实施。
3. 对实验结果进行详细的归纳总结。

Experiment 23 Preparation and Characterization of Drug-loaded Nanoparticles (Comprehensive Designing Experiment)

1. Objective

1.1. To train the ability of students to think independently and to design experiments based on their comprehensive knowledge.

1.2. To familiar with the preparation and characterization of nanoparticles.

1.3. To learn the properties and applications of nanoparticles.

2. Background

Nanoparticles with the size ranging from 1nm to 100nm exhibit special mechanical, electrical, magnetic, optical, and chemical properties due to their small size effect, surface effect and quantum size effect. Nanoparticles have been attracted much attention owing to their unique properties and applications. In recent years, nanotechnology has developed rapidly in the field of medicine. Novel drug preparations such as drug nanocrystals or drug loaded in nanocarriers, are distinctly different from conventional ones in physical and chemical properties, pharmacokinetics and pharmacodynamics. The bioavailability and therapeutic index of drugs are improved due to the sustained-release, controlled-release and targeted features, and the possibility of success in new drug research and developments increased. Nanomedicine has become one of the frontiers and hot topics in the area of pharmaceutical research.

Various nanoparticles such as liposomes, polymeric micelles, nanoparticles and nanocapsulesare are used as nanocarriers to encapsulate drugs, or some water-insoluble drugs can be dispersed to be nanocrystals. The preparation methods of nanoparticles include thin film dispersion method, emulsion solvent evaporation method, high pressure homogeneous method, and colloid mill method, etc. Nanoparticles should have high drug encapsulation efficiency to guarantee its biological effect. In addition, drug loading content should also meet the requirements of drug administration.

The physicochemical properties of drug-loaded nanoparticles are closely related to their biological properties. Therefore, it is necessary to evaluate the physicochemical properties of the obtained drug-loaded nanoparticles. In general, the main properties include drug loading content, encapsulation efficiency, particle size and its dispersion, morphology, and Zeta potential. The particle size and morphology are usually observed under scanning electron microscope (SEM) or transmission electron microscope (TEM). The particle size and its polydispersity, and Zeta potential are determined using dynamic light scattering method.

The students are required to refer to the relevant literatures, to choose appropriate experimental method based on the laboratory condition, to design the experimental procedures, to do experiments and to analyze the data independently.

3. Some suggestions

3.1. Preparation of medicinal nanoparticles

Liposomes are prepared using lecithin and cholesterol. Doxorubicin is loaded into liposomes by pH gradient method or $(NH_4)_2SO_4$ gradient method. Doxorubicin-loaded polymeric micelles are prepared using PEG-PLA by thin film dispersion method. The hydrophilic drug-loaded solid lipid nanoparticles are prepared using glycerol monostearate or polysorbate 80, etc.

3.2. Refer to literatures to understand the preparation technology for drug-loaded nanoparticles and the determination method of drug content in the nanoparticles, as well as the measurement technique of size and morphology. Use the following key words to refer to references: liposomes, nanoparticles, micelles, preparation, characterization.

3.3. Based on the laboratory conditions, select the appropriate preparation method and evaluation method to analyze the composition, morphology, and drug-loading content of the nanoparticles.

4. Apparatus and chemicals

4.1. Apparatus

Rotation evaporator, ultraviolet-visible spectrophotometer, X-ray diffractometer, TEM, SEM, dynamic light scattering particle size analyzer, high speed centrifuge, magnetic stirrer, other commonly used glass instrument.

4.2. Chemicals

Lecithin, cholesterol, PEG-PLA, glycerol monostearate, polysorbate 80, doxorubicin and other reagents.

5. Experimental report

Write the report after finishing the experiments. The report should include the principle of the experiment, experimental method and procedures, data record and processing, results and discussion (the evaluation of experimental design, the influence factors, problems occurring in the experiments and possible reasons, improved vision for the experiment, conclusion, etc.), and references.

6. Requirements

6.1. Use any index tools to refer to relevant literatures.

6.2. According to the references and laboratory condition, design the experimental procedures, do experiments and analyze the data independently.

6.3. Discuss and summarize the resulting detail.

实验二十四 固体药物常规理化常数的测定（综合设计性实验）

（一）实验目的

1. 了解药物常规理化性质和测定方法，培养学生独立实验的能力。

2. 熟悉常用的药物理化性质测定的仪器设备和使用方法。

3. 掌握药物理化性质在药物研究、生产和使用中的意义。

（二）实验背景

药物常规的理化常数有：熔点、沸点、溶解度、吸光系数、比旋度、解离常数、分配系数等，这些理化常数在药物研究、生产和使用过程中都具有十分重要的意义。药物的理化常数是药物质量研究、剂型设计的基础，掌握各种理化常数的意义和测定方法就显得尤为重要。

（三）实验提示

1. 查阅相关文献，特别是各国药典，了解各种常规理化常数的测定方法。

2. 根据实验室条件，设计合理可行的实验方案，与教师研究讨论。

3. 参考以下几点对实验进行设计。

（1）详细写出实验设计的基本思想，完整的实验方法和实验步骤。

（2）原始数据记录表格要根据实验内容自行设计。

（3）实验讨论应包括对实验设计的评价、分析影响实验结果的因素、讨论实验中出现的问题及可能的原因、对实验方案改进的设想等。

（四）实验提供的条件

1. 仪器　差热分析仪，紫外-可见分光光度计，旋光计，电导率仪，熔点管，超级恒温水浴，水浴恒温振荡器，高速离心机，各种常用玻璃仪器。

2. 药品　原料药（10 种以上），各种其他常用试剂。

（五）实验要求

1. 利用各种检索工具查阅相关文献并做出较为详细的摘录。

2. 参考相关文献，通过本人的综合思考，拟订详细的实验方案，并独立实施。

3. 对实验结果进行详细的归纳总结。

4. 以论文形式给出实验报告。

Experiment 24 Determination of Physical and Chemical Constants of Solid Drugs (Comprehensive Designing Experiment)

1. Objective

1.1. To learn the meaning and testing methods of basic physical and chemical constants of drugs. To cultivate the students' independent research skills and ability.

1.2. To familiar with the experimental apparatus and the operation method.

1.3. To understand the significance of the basic physical and chemical constants in drug study, production and application.

2. Background

Basic physical and chemical constants of drugs include melting point, boiling point, solubility, absorptivity, specific rotation, dissociation constant, partition coefficient, and so on, which are of great significance in drug study, production and application. They are related to the quality and formulation designing of the drugs. Therefore, it is much important to study the meanings and testing methods of these constants.

3. Procedures

3.1. Find the information concerned from any possible references, especially from pharmacopoeia of every country. Get familiar with the testing methods of the basic physical and chemical constants.

3.2. Consider the laboratory condition to design a reasonable and feasible experiment.

3.3. Some suggestions:

3.3.1. The report should include the background of the experiment, experimental method and procedures.

3.3.2. Data record charting should be designed according to experiment content.

3.3.3. Experiment discussion should consist of evaluation of experiment design, analysis of experiment result, and improvement for experiment design.

4. Apparatus and chemicals

4.1. Apparatus

Differential thermal analysis (DTA) apparatus, ultraviolet-visible spectrophotometer, polarimeter, electrical conductivity meter, melting point tube, super thermostat, isothermal bath shaker, centrifuge and any other possible apparatus, for example, common glass instruments.

4.2. Chemicals

More than 10 raw material drugs provided, other commonly used regents.

5. Requirements

5.1. Use any index tools to find the information concerned.

5.2. Based on the references, design the experimental procedures under the laboratory condition and carry them out.

5.3. Analyze the data independently and compare them to the result with high standards from references.

5.4. Write the report of experiment in form of thesis.

实验二十五　表面活性剂临界胶束浓度的测定（综合设计性试验）

（一）实验目的

1. 培养学生独立思考、综合运用所学知识设计实验的能力。

2. 熟悉表面活性剂临界胶束浓度的测定方法。

3. 了解表面活性剂的特性及其在药学中的应用。

（二）实验背景

表面活性剂在药学领域中具有广泛的应用。从药剂学、天然药物化学、合成药物化学到药物分析，都涉及表面活性剂，如用作乳化剂、增溶剂、润湿剂、消泡剂、起泡剂、吸收促进剂、杀菌消毒剂等。

表面活性剂分子由疏水基团和亲水基团两部分组成，是两亲性分子。在低浓度水溶液中，表面活性剂主要是以单个分子或者离子的形式存在，定向吸附在水溶液表面。当浓度增加到一定程度时，表面活性剂分子的疏水基通过疏水相互作用缔合在一起，形成疏水基向内、亲水基向外的多分子聚集体，该聚集体称为胶束。形成胶束所需的表面活性剂的最低浓度称为临界胶束浓度（CMC）。CMC 是表面活性剂的一个重要性质参数，它与表面活性剂的性质和功能直接相关。

在 CMC 时，表面活性剂溶液的许多物理性质发生突变。如表面张力、电导率、去污力、增溶作用、渗透压等。因此，可以根据这些物理量随表面活性剂浓度的变化曲线求得 CMC 值。测定 CMC 的方法较多，包括表面张力法、电导率法、染料吸光度法、荧光光度法（以芘作为荧光探针）、滴定法以及光散射法等。在这些方法中，表面张力法、电导率法和荧光分光光度法比较简便、准确。表面张力法灵敏度高，不仅适用于活性较高的表面活性剂，也适用于活性较差的有机物，而且该法不受无机盐的影响，可以测定离子型和非离子型表面活性剂的 CMC。

电导率法是一种经典的测定 CMC 的方法，操作简便，结果可靠。但是，该法只能测定离子型表面活性剂的 CMC，并且过量无机盐的存在会降低其敏感性，因此测定时需要使用电导水配制溶液。

荧光分光光度法是一种测定较低临界胶束浓度（$CMC<1\times10^{-3}mol/L$）的常用方法，这是由于荧光探针芘在胶束内外的荧光振动精细结构不同。该法的优点在于仅需要极少量的芘和表面活性剂。除此之外，该测定方法不受无机盐的影响，既适用于离子型表面活性剂，也适用于非离子型表面活性剂。

本设计实验要求学生自己查阅相关文献，结合实验室的实际情况选择合适的实验方法，自主设计实验方案，独立完成实验操作和数据处理，测定药学中常用的表面活性剂的临界胶束浓度。

（三）实验提示

1. 离子型表面活性剂可选择十二烷基硫酸钠（SDS）、十六烷基三甲基溴化铵（CTAB）等。非离子表面活性剂可选择聚乙二醇辛基苯基醚（TritonX-100）、吐温 80 等。

2. 查阅相关文献，了解表面活性剂的分子结构特点、CMC 值范围、合适的 CMC 测定方法和仪器的使用说明以及 CMC 值的影响因素。文献查阅时使用的关键词（供参考）：表面活性剂、临界胶束浓度。

3. 根据实验室条件选择合适的测定方法，对药用表面活性剂进行临界胶束浓度的测定和分析。

（四）实验室提供的条件

1. 仪器　电导率仪、表面张力测定仪、荧光分光光度计，电磁搅拌器，其他常用玻璃仪器等。

2. 药品　十二烷基硫酸钠，十六烷基三甲基溴化铵，TritonX-100，吐温 80，芘，其他常用和必需试剂。

（五）实验要求

1. 利用各种检索工具查阅相关文献并做出较为详细的摘录。
2. 参考相关文献，通过本人的综合思考，拟订详细的实验方案，并独立实施。
3. 对实验结果进行详细地归纳总结。
4. 以论文形式给出实验报告。

Experiment 25 Determination of Critical Micelle Concentration for Surfactants (Comprehensive Designing Experiment)

1. Objective

1.1. To train the ability of students to think independently and to design experiments based on their comprehensive knowledge.

1.2. To familiar with the determination method of CMC for surfactants.

1.3. To learn the properties and applications of surfactants in the pharmaceutical area.

2. Background

Surfactants have been widely used in the area of pharmaceutics, natural medicine chemistry, chemical synthesis, and pharmaceutical analysis. They are usually used as emulsifier, solubilizer, wetting agent, foaming agent, antifoaming agent, absorption promoter, and sterilization disinfectant.

Surfactant is an amphiphilic substance consisting of both hydrophobic moiety and hydrophilic moiety. When a small amount of surfactant is dissolved in water, it tends to directionally adsorbed on the surface of aqueous solution to reduce the surface tension, forming an oriented mono-molecular membrane. As the surfactant concentration is increased, the hydrophobic group of the surfactant molecule aggregates together through hydrophobic interaction, forming a molecular aggregate which is called "micelle" in which the hydrophobic groups locate inward and the hydrophilic groups are facing to the water. The minimum concentration of surfactant at which micelle begins to form is referred to as the critical micelle concentration (CMC). CMC is an important parameter which is directly related to characteristics and surface activity of surfactants.

Above the CMC, a lot of the physical properties of surfactant solution such as surface tension, conductivity, detergency, solubilization and osmotic pressure are usually changed abruptly. Therefore, it is possible to obtain the CMC value by plots of these physicochemical properties as a function of surfactant concentration. Surface tension method, conductivity method, dye absorbance method, fluorescence method, titration method and light scattering method are commonly used to determine the CMC of surfactants. Among these, surface tension method, conductivity method and fluorescence method are simpler and more precise. Surface tension method with high sensitivities is suitable for both active surfactant and non-active one. Also, the method is not influenced by inorganic salts and therefore suitable for ionic and nonionic surfactants.

Conductivity method, a classic method of determining CMC, is easy to operate and reliable. However, excessive salts will reduce the sensitivity, the solutions should be therefore prepared using conductivity water. The method is only suitable for the ionic surfactants.

Fluorescence method is a common method for measuring CMC of surfactants with very low CMC(typically below 1×10^{-3} mol/L). The advantage of this method is that only a very small amount of fluorescent probe such as pyrene and surfactants are required. Furthermore, this determination method is not affected by inorganic salts, and suitable for both ionic and nonionic surfactants.

The students are required to refer to the relevant literatures, to choose appropriate experimental method based on the laboratory condition, to design the experimental procedures, to do experiments and to analyze the data independently.

3. Some suggestions

3.1. Model surfactants

3.1.1. Ionic surfactants: Sodium dodecylsulfate (SDS), cetyl trimethyl ammonium bromide (CTAB).

3.1.2. Non-ionic surfactant: polyethylene glycol tert-octyl phenyl ether (TritonX-100), Tween 80.

3.2. Refer to literatures to understand the structure characteristics of the surfactants, the range of CMC value, appropriate determination method of CMC, instructions for the instruments, and factors that influence the CMC determination. Use the following key words for references: surfactant, the critical micelle concentration.

3.3. Based on the laboratory conditions, select a suitable method for the determination and analysis of the critical micelle concentration of the surfactants.

4. Apparatus and chemicals

4.1. Apparatus

Conductometer, surface tension meter, fluorescence spectrophotometer, magnetic agitator, commonly used glass instruments.

4.2. Chemicals

Sodium dodecyl sulfate (SDS), cetyl trimethyl ammonium bromide (CTAB), polyethylene glycol tert-octyl phenyl ether(TritonX-100), polysorbate(Tween 80), and other reagents.

5. Requirements

5.1. Use any index tools to refer to relevant literatures.

5.2. According to the references and laboratory condition, design the experimental procedures, do experiments and analyze the data independently.

5.3. Discuss and summarize the results in detail.

5.4. Write the experimental report in the form of a thesis.

实验二十六 乳状液的制备和性质（综合设计性实验）

（一）实验目的

1. 了解乳状液的基本概念。

2. 掌握乳状液的制备以及性质的鉴别方法。

3. 培养学生独立思考、独立设计实验的能力。

4. 了解乳状液的性质在药物研究和使用中的意义。

（二）实验背景

乳剂是一种重要剂型，注射剂、滴眼剂、栓剂、气雾剂、软膏剂都有乳剂型药剂的存在。以乳状液作为载体可制备难溶药物的乳剂，能改善药物的溶解性，具有增加生物利用度，减少毒副作用等优点，具有很大的发展前景。

乳剂的常用制备方法有油中乳化剂法（干胶法）、水中乳化剂法（湿胶法）、新生皂法、两相交替加入法、机械法（乳匀机、胶体磨、超声乳化法）、纳米乳的制备、复合乳剂的制备等。此外，乳剂还大量应用于纳米药物领域，相较于普通剂型，纳米制剂通常可显著提高细胞摄取效率和生物利用度，研究者可以根据纳米制剂的需求来选择不同的乳液制备技术，例如微乳液的超低界面张力、双乳液-溶剂蒸发法的高包封效率以及乳液聚合的抗聚结稳定性等特点，目前乳液技术已成为用于获得高封装效率，高稳定性和低毒性的纳米颗粒的常用方法。

如图 2-26-1 所示，乳状液分油包水型（W/O 型）和水包油型（O/W 型）两种类型，这与形成时所添加的乳化剂性质有关。乳化剂的作用在于降低界面张力，形成一定强度的保护膜，从而使乳状液稳定。常用乳化剂包括油酸钠、失水山梨醇单油酸酯（Span）、聚氧二烯失水山梨醇单油酸酯（Tween）等表面活性剂。此外，当加入某种物质后，乳状液可以由一种类型转变为另一种类型，这种现象称为乳状液的转相。

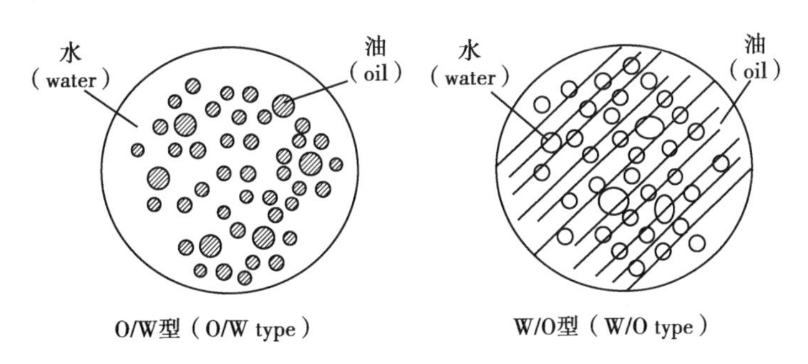

图 2-26-1 乳状液的类型

Fig. 2-26-1 Type of Emulsion

制备乳状液时需要鉴别乳状液的类型，一般采用稀释法、电导法、染色法等。此外，有时需要将乳状液破坏掉，这称之为破乳。常用的破乳方法有电解质破乳、改变乳化剂的类型破乳、破坏保护膜破乳、破坏乳化剂破乳、加热破乳、高压电破乳、离心破乳、过滤破乳等。

本设计要求学生制备并鉴别不同类型的乳状液，并以实验方法证明乳状液的转相和破乳。学生需自己查阅相关文献，结合实验室的实际情况，选择合适的实验方法，自主设计实验方案，独立完成实验操作和数据处理。

（三）实验提示

1. 油相选择环己烷、苯或石油醚（60~90℃）。

2. 查阅相关文献，了解乳状液的性质和特点，了解和借鉴他人研究乳状液的方法。文献查阅时使用的关键词（供参考）：乳状液、制备、性质、转相。

3. 找到合适的方法，制备不同类型的乳状液（水包油、油包水）。

4. 至少选用三种不同方法，鉴别所制备乳状液的类型。以实验方法证明乳状液的转相和破乳。

5. 设计实验方案时,应充分注意实验室的条件。

(四)实验室提供的条件

1. 仪器 电导仪,玻璃棒,具塞锥形瓶,试管,烧杯,量筒,滴定管。

2. 试剂 冰醋酸(AR),正戊醇(AR),环己烷(AR),苯(AR),石油醚(60~90℃),2%油酸钠水溶液,0.4%吐温 80 溶液,1% Span 80 环己烷溶液,1% Span 80 苯溶液,1% Span 80 石油醚溶液,1%苏丹Ⅲ环己烷溶液,0.5%次甲基蓝水溶液,0.2mol/L $CaCl_2$ 溶液。

(五)实验要求

1. 利用各种检索工具查阅相关文献并做出较为详细的摘录。

2. 参考相关文献,通过本人的综合思考,拟订详细的实验方案,并独立实施。

3. 对实验结果进行详细的归纳总结。

4. 以论文形式给出实验报告。

Experiment 26 Preparation and Properties of Emulsion (Comprehensive Designing Experiment)

1. Objective

1.1. To learn the basic concepts of emulsions.

1.2. To learn the preparation of emulsions and the method of property identification.

1.3. To improve the ability of independent thinking and designing experiment.

1.4. To understand the properties of emulsion and their significances in pharmaceutical research and application.

2. Backgrounds

The emulsion is also very important in pharmaceuticals. It can be used to produce many kinds of dosage forms, e.g. oral dosage, local transdermal formulation, injection, etc. The emulsion could be used to improve the bioavailability of drugs with poor solubility. It introduces not only better solubility, but also less side effects, and therefore has been drawn a lot of attention in the research.

The commonly used preparation methods of emulsions include emulsifier in oil (dry glue method), emulsifier in water (wet glue method), new soap method, two-phase alternating addition method, mechanical method (emulsion homogenizer, colloid mill, ultrasonic emulsification method), nanoemulsion preparation, composite emulsion preparation and so on. In addition, the emulsion is also widely used in the field of nanomedicine. Compared with ordinary formulations, nanomaterials can significantly improve cell uptake efficiency and bioavailability. Researchers can choose different emulsion preparation technologies according to the needs of nano preparations. For example, the ultra-low interfacial tension of microemulsions, the high encapsulation efficiency of the double emulsion solvent evaporation method and the anti-coalescence stability of emulsion polymerization are the common methods for obtaining high encapsulation efficiency, high stability and low toxicity nanoparticles.

As shown in Figure 2-26-1, there are two types of emulsion, including water in oil (W/O) and oil in water (O/W). The formation of the types of emulsion depends on the properties of emulsifiers. The emulsifier functions to reduce the interfacial tension between oil and water and to form a coherent monolayer to prevent the coalescence of two droplets when they approach each other. The commonly used emulsifier includes sodium oleate, Span, and Tween. In addition, the emulsion can be changed from W/O type to O/W type with the addition of certain chemicals. This is called the phase transition of emulsion.

In the preparation of emulsion, the identification of emulsion type is necessary. The identification methods usually include dilution, conductivity, and dye methods. Sometimes we need to break the emulsion by the method of demulsification, which in general includes electrolyte demulsification, type-change of emulsifier, destroy of a monolayer, destroy of emulsifier, heating, high voltage, centrifuging and filtration, etc.

In this experiment, the students are required to prepare the emulsions with different types, as well as to identify the phase transition and demulsification of the emulsions. Students should refer to the relevant literatures, to choose appropriate experimental method based on the laboratory condition, to design the experimental procedures, to do experiments and to analyze the data independently.

3. Suggestions

3.1. Oil phase selection. Cyclohexane, benzene or petroleum ether(60~90℃).

3.2. Consulting references to understand the properties and the characteristics of emulsion. Try to understand the methods in the literature and design your own for emulsion preparation. The suggested keywords for literature searching may include emulsion, preparation, properties, and phase transition.

3.3. Prepare different types of emulsions(O/W and W/O types) with appropriate methods.

3.4. Identify the types of emulsions by at least three different methods. Try to prove the phase transition and the demulsification of emulsion with experimental methods.

3.5. Design the experiment with considering the laboratory condition.

4. Apparatus and chemicals

4.1. Apparatus

Conductometer, glass rod, corked conical flask, test tube, beaker, measuring cylinder, dropper.

4.2. Chemicals

Glacial acetic (AR), Pentyl alcoholl. (AR), cyclohexane (AR), benzene (AR), petroleum ether (60~90℃), 2% sodium oleate solution, 0.4% Tween 80 solution, 1% Span 80 in cyclohexane, 1% Span 80 in benzene, 1% Span 80 in petroleum ether, 1% sudan Ⅲ in cyclohexane, 0.5% methylene blue solution, 0.2mol/L $CaCl_2$ solution.

5. Requirements

5.1. Use any literature searching tools to find the needed information.

5.2. Based on the references, design the experimental protocols with detailed procedures. The laboratory conditions shall be considered in the design. Do the experiment and analyze the data independently.

5.3. Discuss and summarize the results in detail.

5.4. Write the report of the experiment in the form of a thesis.

第三部分 实验测试题

一、单 选 题

1. 1mol 液体苯在 298K 时置于氧弹式量热计中完全燃烧，反应式为：$C_6H_6(l)+\frac{15}{2}O_2(g) \longrightarrow 6CO_2(g)+3H_2O(l)$，同时放热 3 264kJ/mol，则其等压燃烧焓为（　　）

A. 3 268kJ/mol　　B. 3 264kJ/mol　　C. −3 264kJ/mol　　D. −3 268kJ/mol

2. 在燃烧焓测定实验中，用雷诺校正曲线求取反应前后温度改变值 ΔT 的主要原因是（　　）

A. 温度变化太快，无法准确读取

B. 系统和环境之间有热交换

C. 消除由于略去有酸形成放出的热而引入的误差

D. 氧弹式量热计绝热，必须校正所测温度值

3. 在燃烧焓测定实验中，点火不成功的可能原因中错误的是（　　）

A. 点火丝太细　　B. 样品受潮

C. 样品压片太紧　　D. 点火电压太小

4. 在燃烧焓测定实验中，为了防止系统漏热，用了下列措施，除了（　　）

A. 准确量取水　　B. 用盖子密封

C. 水桶内表面抛光　　D. 空气层隔热

5. 已知 1% 氯化钠标准溶液的冰点下降为 0.58K，则推算 0.9% 氯化钠注射液的冰点下降为（　　）

A. 0.52K　　B. 0.54K　　C. 0.56K　　D. 0.53K

6. 理想稀溶液的依数性不包括（　　）

A. 凝固点下降　　B. 沸点下降　　C. 渗透压　　D. 蒸气压下降

7. 下列因素中与稀溶液冰点下降值无关的是（　　）

A. 溶剂的种类　　B. 溶剂的数目　　C. 溶质的种类　　D. 溶质的数目

8. 凝固点降低法测定溶质的分子量，若所用的纯溶剂是苯，其正常凝固点为 5.5℃，为使冷却过程在比较接近平衡的情况下进行，比较合适的冷却浴是（　　）

A. 冰-水　　B. 冰-盐水　　C. 干冰-丙酮　　D. 液氮

9. 用凝固点降低法测定溶液的渗透压时，选用的温度计是（　　）

A. 贝克曼温度计　　B. 普通水银玻璃温度计

C. 水银接触温度计　　D. 普通酒精玻璃温度计

10. 含非挥发性溶质的双组分稀溶液的凝固点（T_f）与纯溶剂的凝固点（T_f^*）相比较，正确的是（　　）

A. $T_f>T_f^*$　　B. $T_f<T_f^*$　　C. $T_f=T_f^*$　　D. 不能确定

11. 冰点下降法测定待测样品的渗透压实验中，可能出现冷却太慢，或测不出正确的凝固点的现象，前者和后者产生的原因分别是因为冰浴的温度（　　）

A. 过高和过高　　B. 过高和过低　　C. 过低和过低　　D. 过低和过高

12. 冰点下降法测定待测样品的渗透压实验中，绘制步冷曲线时样品应（　　）

A. 后期快速冷却而前期缓慢冷却　　B. 快速冷却

C. 前期快速冷却而后期缓慢冷却　　D. 缓慢而均匀地冷却

13. 采用冰点下降法测定溶液的渗透压，需准确测定的物理量是（　　）

A. 纯溶剂的凝固点　　B. 溶液的凝固点

C. 溶液和纯溶剂凝固点的差值　　D. 可溶性溶质的凝固点

14. 采用凝固点下降法测定溶液的渗透压时，下列方法中不能有效降低过冷程度的是（　　）

A. 快速搅拌待测溶液　　B. 测定管外加入空气夹套

C. 向溶液内加入晶种　　D. 加热溶液

15. 在动态法测定水的饱和蒸气压实验中，实验温度在 80~100℃之间，则所测得的气化热数据为（　　）

A. 水在 80℃时的气化热　　B. 水在 100℃时的气化热

C. 实验温度范围内气化热的平均值　　D. 该数值与温度无关

16. 在测定纯水的饱和蒸气压的实验中，若通过测定不同外压下纯水的沸点来进行，则这种测定方法属于（　　）

A. 静态法　　B. 动态法　　C. 饱和气流法　　D. 流动法

17. 在静态法测定纯水的饱和蒸气压的实验中，正确的调压进气操作是（　　）

A. 必须在空气排净和恒温后可调压　　B. 边加热边调压

C. 等气泡冒完后即可调压　　D. 只在恒温后可调压

18. 分配系数的主要影响因素是（　　）

A. 温度　　B. 浓度　　C. 体积　　D. 大气压

19. 在分配系数测定实验中，若滴定水层时过量，对分配系数的影响为（　　）

A. 偏大　　B. 偏小　　C. 无影响　　D. 不确定

20. 在分配系数测定实验中，滴定过程的溶液颜色变化为（　　）

A. 淡黄→蓝色→红色　　B. 淡黄→无色→蓝色

C. 淡黄→绿色→无色　　D. 淡黄→蓝色→无色

21. 在测定碘和碘化钾平衡常数实验中，用 $Na_2S_2O_3$ 标准溶液滴定时，四氯化碳层溶液的终点颜色和水层的终点颜色分别为（　　）

A. 无色和淡蓝色　　B. 淡黄色和无色

C. 无色和淡黄色　　D. 淡红色和淡蓝色

22. 在测定碘和碘化钾平衡常数实验中，影响平衡常数测定的实验条件主要有（　　）

A. 温度　　B. 反应体积

C. 滴定 CCl_4 层时加入的 KI　　D. 指示剂浓度

23. 二组分部分互溶双液系统相图中，会溶温度越高说明(　　)

A. 两种液体相互溶解的能力越强

B. 两种液体相互溶解的能力越差

C. 不能说明两种液体的相互溶解的能力

D. 以上说法都不正确

24. 二组分部分互溶双液系统相图实验中，绘制的苯酚-水的温度-组成相图具有(　　)

A. 最低会溶温度　　B. 最高会溶温度

C. 既有最高又有最低会溶温度　　D. 既无最高也无最低会溶温度

25. 溶解度法绘制苯酚-水的温度-组成相图实验中，没有用到的实验仪器是(　　)

A. 贝克曼温度计　　B. 水浴　　C. 磁力加热搅拌器　　D. 空气套管

26. 下列系统中可以用溶解度法绘制相图的是(　　)

a. 苯酚-水　b. 环已烷-乙醇　　c. 水-盐　　d. 萘-苯

A. a,b　　B. b,d　　C. a,c　　D. a,d

27. 在双液系统气液平衡相图绘制实验中，常选择测定物系的折光率来测定物系的组成。下列说法中不属于选择该方法的理由的是(　　)

A. 折光率测定操作简单　　B. 该方法对任何双液系统都能适用

C. 测定所需的试样量少　　D. 测量所需时间少，速度快

28. 将含环已烷 30% 的环已烷-乙醇溶液在 101.325kPa 进行精馏时，如果塔效率足够，则下列说法正确的是(　　)

A. 塔顶馏出液为环已烷　　B. 塔釜残留液为环已烷

C. 塔顶馏出液为乙醇　　D. 塔釜残留液为乙醇

29. 若步冷曲线上的平台内侧有凹陷，则表示冷却过程中出现了(　　)

A. 过热现象　　B. 过饱和现象　　C. 过冷现象　　D. 化学反应

30. 低共熔组成的合金或纯金属都只有一个平台转折点，则低共熔组成的合金的熔点(T)与纯金属的熔点(T^*)的关系为(　　)

A. $T< T^*$　　B. $T=T^*$　　C. $T>T^*$　　D. 不能确定

31. 当用三角形坐标来表示三组分物系时，若某物系其组成在平行于底边 BC 的直线上变动时，则该物系的特点是(　　)

A. B 的百分含量不变　　B. A 的百分含量不变

C. C 的百分含量不变　　D. B 和 C 的百分含量之比不变

32. 关于氯仿/乙酸/水三元相图绘制实验中用到的玻璃器皿，下列说法正确的是(　　)

A. 用蒸馏水润洗后，可直接使用

B. 需要干燥

C. 实验前应该用待测液仔细润洗

D. 实验前应分别称取其质量，并记录

33. 关于电导率测定和应用实验，下列说法正确的是(　　)

A. 氢氧化钠滴定醋酸时，电导率值先降后升

B. 测定醋酸解离常数时，需同时测得醋酸溶液和纯水的电导率

C. 测定氯化银的水中溶解度,需同时测得氯化银饱和溶液和纯水的电导率

D. 醋酸和醋酸钠的 Λ_m^∞ 都可以通过线性外推法求得

34. 测纯水电导率时,若在空气中放置了一段时间,则测得值将(　　)

A. 增大　　B. 减小　　C. 没有变化　　D. 无法确定

35. 测量溶液的电导时,使用的电极是(　　)

A. 铂黑电极　　B. 玻璃电极

C. 甘汞电极　　D. 银-氯化银电极

36. 电导率仪在测量数值之前,需先进行(　　)

A. 零点校正　　B. 确定电导池常数

C. 满刻度校正　　D. 以上都需进行

37. 可逆电池电动势必须用电位差计来测量,主要是为了(　　)

A. 减少标准电池的损耗　　B. 其测量原理为对消法,接近可逆条件

C. 消除电极上的副反应　　D. 简单易测

38. 在用电位差计测电池电动势时,首先要接通标准电池进行"标准化"操作,其目的是(　　)

A. 检查线路连接的正确性　　B. 校正检流计零点

C. 校正标准电池电动势　　D. 标定工作电阻

39. 对于同一个样品,采用同一台旋光仪,在测定其旋光度时,分别用长度为 10cm 和 20cm 旋光管进行测量,所测定的旋光度分别为 α_1 和 α_2,则(　　)

A. $2\alpha_1=\alpha_2$　　B. $\alpha_1=\alpha_2$　　C. $\alpha_1=2\alpha_2$　　D. 不确定

40. WZZ-2 型自动旋光仪采用的光源是(　　)

A. 蓝光　　B. 白光　　C. 钠光灯　　D. 红光

41. 在蔗糖水解反应实验中,下列操作正确的是(　　)

A. 将蔗糖与盐酸分别盛放在两个锥形瓶中,然后同时倒入一个烧杯

B. 用移液管移取盐酸注入盛有蔗糖的锥形瓶

C. 用移液管移取蔗糖注入盛有盐酸的锥形瓶

D. 将盛放在锥形瓶中的盐酸迅速倒入另一盛有蔗糖的锥形瓶中

42. 关于蔗糖水解反应,下列说法正确的是(　　)

A. 是一级反应,水不参与化学反应

B. 是假一级反应,水在反应前后浓度变化不大

C. H^+不参与化学反应

D. H^+在反应前后浓度变化不大

43. 在丙酮溴化反应速率常数的测量实验中,为了方便、准确地测量反应进程,最合适的选用仪器是(　　)

A. 电导率仪　　B. 阿贝折光仪　　C. 分光光度计　　D. 旋光仪

44. 在丙酮溴化反应速率常数的测定实验中,用实验测得的吸光度计算所得的浓度是(　　)

A. c_{Br_2}　　B. c_{H^+}　　C. $c_{丙酮}$　　D. 以上都不是

45. 在丙酮溴化反应速率常数的测定实验中,反应速率的测定方法为(　　)

A. 半衰期法　　B. 积分法　　C. 孤立法　　D. 初速率法

46. 在乙酸乙酯皂化反应的动力学实验中,采用物理法测定乙酸乙酯的浓度变化,使用的测量仪器是(　　)

A. 旋光仪　　B. 电导率仪　　C. 阿贝折光仪　　D. 沸点仪

47. 在乙酸乙酯皂化反应的动力学实验中,若将氢氧化钠加入乙酸乙酯中一半时作为反应起点,不考虑酯的挥发,则所测结果(　　)

A. 有正误差　　B. 有负误差　　C. 无误差　　D. 误差不可测

48. 在乙酸乙酯皂化反应的动力学实验中,若乙酸乙酯溶液提前配好,将导致反应过程中电导值(　　)

A. 增大　　B. 减小　　C. 不变　　D. 不确定

49. 在乙酸乙酯皂化反应中,系统的电导率值随时间的变化情况是(　　)

A. 增大　　B. 减小　　C. 不变　　D. 先小后大

50. 在碘化钾与过氧化氢反应的速率常数及活化能的测定实验中,需要准确加入的是(　　)

A. KI 溶液　　B. 淀粉溶液　　C. $Na_2S_2O_3$ 溶液　　D. H_2SO_4 溶液

51. 关于碘化钾与过氧化氢反应的速率常数及活化能的测定实验,下列说法正确的是(　　)

A. 若最初 1ml $Na_2S_2O_3$ 溶液消耗完后没有及时计时,实验即告失败,必须重做

B. 实验过程中出现蓝色时,立即先加入 $Na_2S_2O_3$ 溶液,然后再计时

C. 若第二次较高温度的实验速率过快,可通过减小 $Na_2S_2O_3$ 溶液的浓度来减小反应速率

D. 在配制 KI 溶液时,若有游离 I_2 而使淀粉变蓝,可滴加几滴 $Na_2S_2O_3$ 消去蓝色

52. 在最大泡压法测定溶液表面张力的实验中,如果将毛细管底端插入待测溶液中,则所得结果将(　　)

A. 偏大　　B. 偏小　　C. 不变　　D. 不能确定

53. 在测定溶液表面张力实验中,准确测定异戊醇(或正丁醇)浓度的仪器是(　　)

A. 电导率仪　　B. 阿贝折光仪　　C. 分光光度计　　D. 旋光仪

54. 在最大泡压法测定溶液表面张力的实验中,下列实验操作不正确的是(　　)

A. 毛细管壁必须严格清洗干净

B. 毛细管应垂直放置,其底端管口刚好与液面相切

C. 毛细管底端管口必须平整

D. 毛细管垂直插入液体内部,且每次插入深度尽量保持不变

55. 在最大泡压法测定液体表面张力的实验中,从毛细管中逸出的气泡的曲率半径 r 的变化过程是(　　)

A. 逐渐变大　　B. 先逐渐变大再逐渐变小

C. 逐渐变小　　D. 先逐渐变小再逐渐变大

56. 固体在溶液中的吸附实验中,对滴定用的锥形瓶作如下处理:①洗涤后锥形瓶未完全干燥;②用待测液洗涤锥形瓶两次。这些操作对实验结果的影响是(　　)

A. ①没影响,②有影响　　B. ①、②都没影响

C. ①、②都有影响　　D. ①有影响,②没影响

57. 根据兰格缪尔单分子层吸附模型计算的固体比表面与实际值相比(　　)

A. 偏大　　B. 偏小
C. 基本相等　　D. 误差太大,没意义

58. 在固体自溶液中的吸附实验中,从锥形瓶里取样分析时,若不小心取了少量的活性炭,对实验结果的影响是(　　)
A. 无影响
B. 使测得的平衡浓度偏大,计算的吸附量将偏大
C. 使测得的平衡浓度偏小,计算的吸附量将偏大
D. 使测得的平衡浓度偏大,计算的吸附量将偏小

59. 下列分散系统中丁铎尔效应最强的是(　　)
A. 空气　　B. 蔗糖水溶液　　C. 大分子溶液　　D. 硅胶溶胶

60. 用0.08mol/L的KI和0.1mol/L的$AgNO_3$溶液以等体积混合制成的溶胶,相同浓度的$CaCl_2$、Na_2SO_4、$MgSO_4$溶液对它的聚沉能力顺序为(　　)
A. $Na_2SO_4>CaCl_2>MgSO_4$　　B. $MgSO_4>Na_2SO_4>CaCl_2$
C. $Na_2SO_4>MgSO_4>CaCl_2$　　D. $Na_2SO_4< MgSO_4< CaCl_2$

61. 溶胶的ζ电势的测定实验中,电泳管中加入的盐酸是(　　)
A. 浓盐酸　　B. 浓度为0.1mol/L的稀盐酸
C. 电导率与溶胶相同的稀盐酸　　D. 以上都可以

62. 在溶胶的制备实验中,$Fe(OH)_3$溶胶的制备方法是(　　)
A. 改换溶剂法　　B. 胶溶法　　C. 水解法　　D. 以上都不是

63. 在黏度法测定大分子的分子量实验中,直接测定的物理量是(　　)
A. 时间　　B. 相对黏度　　C. 增比黏度　　D. 特性黏度

64. 在测定大分子的黏度实验中,恒温槽中使用的介质为(　　)
A. 硅油　　B. 甘油　　C. 水　　D. 待测液

65. 关于黏度法测定大分子的分子量实验,下列叙述错误的是(　　)
A. 用毛细管黏度计测定大分子的黏度,所需主要仪器设备是毛细管黏度计、恒温系统、秒表
B. 测定液体黏度的黏度计类型有毛细管、落球、旋转(或扭力)黏度计
C. 毛细管法测定大分子的黏度实验中,直接测定的物理量是大分子溶液流过毛细管的时间、恒温槽温度
D. 用毛细管法测定液体黏度时,不必使用恒温系统

二、简 答 题

1. 在燃烧焓实验中,为了防止系统漏热,可采取哪些措施?
2. 简述氧弹式量热计测定物质燃烧焓的主要误差来源。
3. 电热补偿法测定溶解热需要使用哪些仪器?
4. 物质溶解时的热效应取决于哪些因素?
5. 已知某物质的分子质量为146,它溶于环己烷中,但不知道它在环己烷中的存在状态(单体还是二聚体,或两者的平衡状态),试设计实验,根据实验结果指出具体存在状态。
6. 在静态法测定液体饱和蒸气压实验中,能否在加热条件下检查系统的密闭性?为

什么？

7. 在反应平衡常数及分配系数的测定实验中，所用的锥形瓶哪些需要干燥？哪些不需要干燥？为什么？

8. 在反应平衡常数及分配系数的测定实验中，需要直接测得哪些实验数据？如何测得？

9. 在反应平衡常数及分配系数的测定实验中，装好溶液后的锥形瓶为什么要立即塞紧磨口塞？

10. 在完全互溶双液系统相图的绘制实验中，如何判断气液两相已达平衡？

11. 在完全互溶双液系统相图的绘制实验中，要注意的问题有哪些？

12. 三组分液-液体系的平衡相图实验中，要绘制单相区与两相区的分界线（即双结点溶解度曲线或双结线），应准确记录哪些数据，知道哪些数据，计算出哪些数据？

13. 使用电导率仪时应注意哪些问题？

14. 在电动势法测定电池反应的热力学函数实验中，化学反应为 $Ag(s)+\frac{1}{2}Hg_2Cl_2(s)\longrightarrow AgCl(s)+Hg(l)$，设计的电池为 $Ag(s)|AgCl(s)|KCl(0.1mol/L)|Hg_2Cl_2(s)|Hg(l)$。为什么这里选用 0.1mol/L 的 KCl 而不是高浓度的 KCl 做电极液？

15. 在旋光法测定蔗糖转化反应的速率常数实验中，在混合蔗糖溶液和 HCl 溶液时，我们将 HCl 液加到蔗糖溶液中。那么反过来混合可以吗？

16. 为什么电导法测定乙酸乙酯皂化反应的速率常数实验中要在恒温条件下进行，且溶液在混合前还要预先恒温？

17. 在碘化钾与过氧化氢反应的速率常数测定实验中，溶液体系的酸碱性对实验有何影响？

18. 影响丙酮溴化反应速率常数测定结果的主要因素是什么？

19. 影响表面张力测定结果的因素有哪些？如何减少或消除这些因素对实验结果的影响？

20. 在固体在溶液中的吸附实验中，如何判断吸附达到平衡状态？

21. 溶胶形成的基本条件是什么？

22. 电泳速度的快慢与哪些因素有关？

23. 在 $Fe(OH)_3$ 溶胶 ζ 电势的测定实验中，选用了稀 HCl 为辅助液。试回答辅助液的选取有什么要求？

24. 在溶胶的 ζ 电势的测定实验中，为什么辅助液的电导率需尽可能与溶胶的相等？

25. 乌贝路德黏度计中支管 C 有何作用？除去支管 C 是否仍可测定黏度？

26. 乌贝路德黏度计的毛细管太粗或太细对测定结果有何影响？

27. 为什么用特性黏度[η]来求算大分子的平均相对分子质量？它和纯溶剂黏度有区别吗？

三、答　　案

一、单选题

1. D　2. B　3. A　4. A　5. D　6. B　7. C　8. A　9. A
10. B　11. B　12. D　13. B　14. D　15. C　16. B　17. A　18. A

19. B	20. D	21. C	22. A	23. B	24. B	25. A	26. C	27. B
28. D	29. C	30. A	31. B	32. B	33. C	34. A	35. A	36. D
37. B	38. D	39. A	40. C	41. D	42. B	43. C	44. A	45. D
46. B	47. B	48. A	49. B	50. C	51. D	52. A	53. B	54. D
55. C	56. A	57. B	58. D	59. D	60. C	61. C	62. C	63. A
64. C	65. D							

二、简答题

1. 答:可采取的措施为:①用盖子密封,②水桶内表面抛光,③空气层隔热。

2. 答:引起误差的主要因素包括:①样品称量;②系统与环境之间的漏热;③燃烧完全程度;④两次实验所用的水量不等;⑤搅拌引起的额外热效应;⑥空气中 N_2 变成硝酸所引起的热效应;⑦升温系统仪表的灵敏度等。

3. 答:需要使用直流电源,数字电压表(精度 0.001V),数字电流表(精度 0.001A),量热计(包括保温杯和加热器),磁力搅拌器,数字贝克曼温度计,压片机,计时表,分析天平,台式天平。

4. 答:溶质溶解于溶剂的过程由溶质晶格破坏、电离的吸热过程和溶质溶剂化的放热过程组成,总的热效应取决于两者之和,可能是吸热的,也可能是放热的。物质溶解时的热效应与压力、温度、溶剂种类以及溶质和溶剂的相对量有关。

5. 答:可以用凝固点降低测定相对分子质量(或摩尔质量)的方法,计算公式为:$M_B = K_f W_B / \Delta T_f W_A$。若环己烷的凝固点降低常数 K_f 已知,通过测定纯环己烷的凝固点(T^*),以及已知溶剂质量 W_A 和溶质质量 W_B 的溶液的凝固点(T)得到 ΔT_f,代入上式计算 M_B。若 M_B 算得为 146,则为单体存在,若为 292,则为二聚体存在,若在 146~292 之间,则为两者平衡存在。

6. 答:不能,因为加热过程中温度不能恒定,气-液两相不能达到平衡,压力也不恒定,造成所测结果不准确。

7. 答:①用标准 $Na_2S_2O_3$ 滴定时所用的锥形瓶,装碘的饱和水溶液的锥形瓶不需要干燥,因为其对碘的物质的量没有影响,不会引起偏差;②装 KI 溶液的碘量瓶需要干燥,若有水会影响碘离子的浓度。

8. 答:①恒温水浴温度,水层和 CCl_4 层中碘的浓度。②水浴温度用普通温度计测量,碘的浓度用硫代硫酸钠标准溶液滴定得到。

9. 答:因为碘和四氯化碳易挥发,立即塞紧磨口塞是为了防止碘和四氯化碳挥发。

10. 答:温度计的读数基本不再变化,或者气相和液相的组成基本不随时间变化而变化(即折光率的测得值基本不再改变),即可认定气液两相基本达到平衡。

11. 答:①为保证系统内外气压的一致,系统不能密闭;②测定样品的折光率时,先测定气相冷凝液,再测定液相冷凝液;③加热棒应浸入溶液 2cm 左右;④使用阿贝折光仪时必须注意保护棱镜,滴管尖不能碰触镜面。如果需要,用拭镜纸轻拭镜面,不能用滤纸擦拭。

12. 答:①应准确记录清浊转变时各组分的体积,实验温度;②应知道实验温度下各纯组分的密度;③计算出各组分的质量及质量百分数,然后将相应数据绘制于等边三角形坐标系中即可得到。

13. 答:①电极引线不能潮湿,否则所测数据不准;②纯水的测量要迅速,否则会因二氧化碳溶解在水中,改变水的电导率,影响测量结果;③盛待测液的容器必须清洁,无离子污染;④擦拭电极时不可触及铂黑,以免影响电极常数。

14. 答:虽然该电池的电动势与 KCl 溶液的活度无关,即 $E=E^{\ominus}$,但是,KCl 浓度太大对银-氯化银电极有溶解作用,故本实验采用 0.1mol/L 的 KCl 作电极液。

15. 答:不能,因为 H^+是催化剂,将反应物蔗糖加入到大量 HCl 溶液时,H^+浓度很大,一旦加入马上分解产生果糖和葡萄糖,则在开始测量时,已经有大部分蔗糖产生了反应,记录 t 时刻对应的旋光度已经不再准确。反之,将 HCl 溶液添加到蔗糖溶液中去,由于 H^+的浓度小,反应速率小,计时之前所进行的反应的量很小。

16. 答:由于反应速率常数以及电导率的数值都与温度有关,要测得准确的动力学数据,必须在恒温条件下进行。预先恒温可以减少混合时温度的波动,减少实验误差。

17. 答:本实验测定的反应为:$2H^+ + 2I^- + H_2O_2 \longrightarrow 2H_2O + I_2$。如果在中性和碱性条件下,过氧化氢被碘化钾催化分解为氧气;而在酸性条件下,过氧化氢被还原为水,所以在本实验中一定要保证是酸性条件。

18. 答:影响本实验结果的主要因素有:溶液浓度、波长、温度、反应时间。

19. 答:温度、气泡逸出速度、毛细管的洁净情况以及毛细管的底端与液面的相对位置等因素会影响表面张力的测定结果。减小或消除这些因素引起误差的措施是:待测液保持恒温,控制气泡逸出速度,毛细管清洗干净,毛细管与液面相切。

20. 答:在不同时刻取等量的少许溶液滴定,若滴定消耗的 NaOH 溶液的体积不随时间发生变化,则可判定吸附已经达到了平衡。

21. 答:形成溶胶的必要条件:①分散相的溶解度要小;②还必须有稳定剂存在,否则胶粒易聚结而聚沉。

22. 答:由 ζ 电势的计算公式 $\zeta=9\times10^9 \cdot \frac{6\pi\eta v}{\varepsilon_r E}$,可得 $v=\frac{1}{9\times10^9} \cdot \frac{\zeta\varepsilon_r E}{6\pi\eta}$。由此式可知,电泳速度的快慢与溶胶的 ζ 电势、介质的介电常数 ε_r、电泳仪两极间的电场强度 E 成正比,与溶胶的黏度 η 成反比。

23. 答:①与溶胶无化学反应;②不能使溶胶发生聚沉;③能与溶胶形成清晰的界面;④和溶胶的电导尽可能相等。

24. 答:①保证辅助液的移动速度与溶胶相等,避免电场强度在界面处的突变造成两管界面移动速度不等而产生界面模糊。②只有电导率相近,计算公式才成立。

25. 答:支管 C 的作用是与大气相通,使得 B 管中的液体完全是靠重力落下而不受其他因素的影响。除去支管 C 仍可测定黏度,只不过此时标准溶液和待测溶液体积必须相同,因为此时的液体下流时所受的压力差与 A 管中液面高度有关。

26. 答:毛细管太粗流经时间短,容易使读数误差增大。毛细管太细容易造成堵塞。

27. 答:①特性黏度定义为当高分子溶液浓度趋于零时的比浓黏度,表示单个分子对溶液黏度的贡献,其值与浓度无关,但与高分子的相对分子质量存在着定量关系,$[\eta]=KM_\eta^\alpha$。所以,常用特性黏度的数值来求取大分子的相对分子质量。②特性黏度与纯溶剂的黏度不同,$[\eta]$主要反映的是无限稀释溶液中共聚物分子与溶剂分子之间的内摩擦作用,纯溶剂黏度反映的是溶剂分子之间的内摩擦作用。

Part Three Experiment Test Questions

Section One Multiple Choice

1. One mole of liquid benzene undergoes complete combustion in a bomb calorimeter at 298K, the reaction is: $C_6H_6(l)+\frac{15}{2}O_2(g)\longrightarrow 6CO_2(g)+3H_2O(l)$ the heat released is 3 264 kJ/mol, then its enthalpy of combustion at constant pressure is ( )

A. 3 268kJ/mol B. 3 264kJ/mol C. −3 264kJ/mol D. −3268kJ/mol

2. In the experiment to determine the enthalpy of combustion, the Reynolds correction curve is employed to obtain the ΔT precisely, because ( )

A. the change of temperature is too fast to obtain accurately

B. there is heat transfer between systems and surroundings

C. it needs to eliminate errors in omitting the heat released when acid substances are formed

D. the temperature value must be corrected due to adiabatic bomb calorimeter employed

3. In the experiment to determine the enthalpy of combustion, ignition fails due to the following reasons except for ( )

A. The igniter fuse is too thin B. The sample is damp

C. The tablet of samples is too hard D. The ignition voltage is too low

4. In the experiment to determine the enthalpy of combustion, in order to avoid the heat leakage from the systems, the following methods can be used except ( )

A. to take the water precisely

B. to seal the cover tightly

C. to polish the inner surface of water bucket

D. to use air-layer for heat insulation

5. Given that the freezing point depression of 1% of sodium chloride standard solution was 0.58K, then the freezing point depression estimated for 0.9% sodium chloride injection is ( )

A. 0.52K B. 0.54K C. 0.56K D. 0.53K

6. Which of the following properties does not belong to the colligative properties of ideal dilute solutions? ( )

A. freezing point depression B. boiling point depression

C. osmotic pressure D. vapor pressure depression

7. Among the following factors, the one that shows no relevant to the freezing point depression of diluted solutions is ( )

A. the type of solvent B. the amount of solvent

C. the type of solute D. the amount of solute

8. In the experiment of determination of molecular weight of solute by freezing point depression method, if the pure solvent is benzene, its normal freezing point is 5.5℃. Then the appropriate cooling bath being used to make the cooling process relatively close to balance state is ( )

A. ice-water B. ice-salt C. dry ice-acetone D. liquid nitrogen

9. In the experiment of determination of osmotic pressure of solution by freezing point depression method, the thermometer used is ( )

A. Beckmann thermometer B. common mercury-in-glass thermometers

C. mercury contact thermometer D. common alcohol-in-glass thermometers

10. The freezing point (T_f) of the two-component dilute solution containing nonvolatile solutes is compared with the freezing point (T_f^*) of the pure solvent. The correct one is ( )

A. $T_f>T_f^*$ B. $T_f<T_f^*$ C. $T_f=T_f^*$ D. uncertain

11. In determination of osmotic pressure of solutions by freezing point depression method, the phenomenon of too slow cooling rate or inaccurate freezing point determination might appear. The explanation for the phenomenon is ( )

A. overhigh bath temperature for both

B. overhigh and overlow bath temperature for former and latter respectively

C. overlow bath temperature for both

D. overlow and overhigh bath temperature for former and latter respectively

12. To draw the cooling curve in the determination of the osmotic pressure by freezing point depression method, the sample should be ( )

A. cooled fast later but slowly first B. cooled fast

C. cooled fast first but slow down later D. cooled slowly and evenly

13. To determine the osmotic pressure of the sample by method of freezing point depression, one should accurately determine ( )

A. the freezing point of pure solvent

B. the freezing point of solution

C. the difference of the freezing point between solution and pure solvent

D. the freezing point of soluble solute

14. To determine the osmotic pressure of the solution by freezing point depression method, the approach that fails to reduce supercooling effectively is ( )

A. fast stirring the solution

B. providing the air jacket outside the determining tube

C. addition of seed crystals to the solution

D. heating the solution

15. In the experiment of determination of saturated vapor pressure of water by dynamic method, the experimental temperature is between 80℃ and 100℃. Then the measured heat of vaporization is ( )

A. the value at 80℃

B. the value at 100℃

C. the average value in the range of experimental temperature

D. the value is independent of temperature

16. Supposing that the saturated vapor pressure of pure water is measured by determination of boiling point of pure water under different external pressures, this method belongs to ( )

A. static method B. dynamic method

C. saturated gas flow method D. flow method

17. To use the static method to measure the saturated vapor pressure of the liquid, the correct procedure for adjusting the pressure of air intake is ( )

A. to adjust pressure after net emission of air and constant temperature

B. to adjust pressure while heating

C. to adjust pressure only after no bubbling

D. to adjust pressure only after constant temperature

18. The main factor that influences the partition coefficient is ( )

A. temperature B. concentration

C. volume D. barometric pressure

19. To determine the partition coefficient, if the titration of aqueous layer is excessive, the partition coefficient determined will be ( )

A. larger than actual B. smaller than actual

C. no influence D. uncertain

20. In the experiment of determination of partition coefficient, the correct color change of the solution in the process of titration is ( )

A. faint yellow→blue→red B. faint yellow→colorless→blue

C. faint yellow→green→colorless D. faint yellow→blue→colorless

21. In the experiment of determination of equilibrium constant for the reaction of I_2 with KI, the colors of end point during titration with $Na_2S_2O_3$ standard solution for both the CCl_4 layer and aqueous layer are respectively ( )

A. colorless and faint blue B. faint yellow and colorless

C. colorless and faint yellow D. faint red and faint blue

22. In the experiment of determination of equilibrium constant for the reaction of I_2 with KI, the main factor that influences the determination is ( )

A. temperature B. reaction volume

C. KI added into CCl_4 layer during titration D. indicator concentration

23. In the phase diagram of partially miscible binary liquid system, the higher critical solution temperature means that ( )

A. two liquids have higher miscibility B. two liquids have lower miscibility

C. it cannot tell the miscibility D. none of the above is correct

24. The temperature-constitutional phase diagram for partially miscible system of phenol and water under constant pressure has ( )

A. minimum critical solution temperature

B. maximum critical solution temperature

C. both maximum and minimum critical solution temperature

D. neither maximum nor minimum critical solution temperature

25. In the experiment of constructing the temperature-constitutional diagram for phenol and water under constant pressure by solubility method, the experimental instrument unused is ( )

A. Beckmann thermometer B. water bath

C. magnetic stirrer with heating D. air casing

26. Among the following systems, the one that its phase diagram can be drawn by solubility method is ( )

a. phenol-water b. cyclohexane-ethanol c. water-salt d. naphthalene-benzene

A. a, b B. b, d C. a, c D. a, d

27. In the experiment of gas-liquid equilibrium for miscible binary liquid system, the refractive index is measured to determine the composition of the system. Among the following statements, the one that is not the reason to choose this method is ( )

A. The operation of measuring refractive index is simple

B. It can be applied to any two-liquid system

C. The amount of sample required for determination is small

D. The measuring time is less and the speed is fast

28. Supposing the tower efficiency is sufficient, which of the following statements is correct about distilling a cyclohexane-ethanol solution containing 30% cyclohexane at 101.325kPa? ( )

A. The distillate at the top of the tower is cyclohexane solution

B. The residual liquid in the tower kettle is cyclohexane solution

C. The distillate at the top of the tower is ethanol solution

D. The residual liquid in the tower kettle is ethanol solution

29. If there was a depression in the inner side of the platform of the step cooling curve, which of the following phenomena appeared during the cooling process? ( )

A. superheating B. supersaturation

C. supercooling D. chemical reaction

30. There is only one platform turning point in the step cooling curve for low eutectic alloy or pure metal. Then the relationship between the melting point of eutectic alloy (T) and the melting point of pure metal (T^*) is ( )

A. $T<T^*$ B. $T=T^*$ C. $T>T^*$ D. uncertain

31. The three-component system is represented by triangular coordinate. If the system point moves along a straight line parallel to bottom line BC, then ( )

A. the percentage of component B keeps unchanged

B. the percentage of component A keeps unchanged

C. the percentage of component C keeps unchanged

D. the ratio of percentages of component B to component C keeps unchanged

32. Which of the following statements is correct about the glassware used in the experiment of drawing the ternary phase diagram for system of chloroform, acetic acid and water at constant temperature and pressure? (　　)

A. The glassware can be used directly after being rinsed with distilled water.

B. The glassware needs to be dried.

C. The glassware needs to be carefully rinsed with the measuring solution before the experiment.

D. The glassware needs to be weighed and recorded before the experiment.

33. Which of the following statements is correct about determination and application of conductivity? (　　)

A. When titrating the HAc solution with NaOH solution, the conductivity decreases firstly and then increases.

B. To determine the dissociation constant of HAc, both the conductivities of HAc solution and pure water need to be measured.

C. To determine the solubility of AgCl, both the conductivities of AgCl saturated solution and pure water need to be measured.

D. The Λ_m^∞ of both the HAc and NaAc can be obtained by linear extrapolation method.

34. When determining the conductivity of pure water, if set the sample in the air for a while, then the conductivity value will be (　　)

A. increased　　B. decreased　　C. unchanged　　D. uncertain

35. When determining the conductivity of solution, the electrode used is (　　)

A. platinum black electrode　　B. glass electrode

C. calomel electrode　　D. silver-silver chloride electrode

36. When using the conductometer to determine the conductivity, the operation need to be done in advance is (　　)

A. zero check　　B. cell constant choose

C. full scale check　　D. all of the above

37. The EMF of a reversible cell can be determined with a potentiometer because (　　)

A. it can reduce loss of the standard cell

B. it employs the compensation technique being close to the reversible condition

C. it can eliminate the side reaction on electrode

D. the determination procedures are simple

38. To measure the EMF using a potentiometer, setting the switch first to standard cell and performing standardized operation are required. The purpose of doing so is (　　)

A. to ensure a correct circuit connection

B. to perform the zero calibration of galvanometer

C. to calibrate the EMF of standard cell

D. to demarcate the working resistor

39. For the same sample, using the same polarimeter to determine its rotation, the polarizers are 10cm and 20cm in length respectively, the optical rotations are α_1 and α_2 respectively. Then the relationship between α_1 and α_2 is ( )

A. $2\alpha_1=\alpha_2$ B. $\alpha_1=\alpha_2$ C. $\alpha_1=2\alpha_2$ D. uncertain

40. The light source used in WZZ-2 type automatic polarimeter is ( )

A. blue light B. white light C. natrium lamp D. red light

41. Which of the following statements is correct about operation in the experiment of hydrolytic reaction of sucrose? ( )

A. The sucrose and hydrochloric acid are placed in two conical flasks and added into a beaker at the same time.

B. Use pipette to take hydrochloride acid and add it into conical flask which contains sucrose.

C. Use pipette to take sucrose and add it into conical flask which contains hydrochloride acid.

D. The hydrochloric acid in conical flasks is added quickly into another conical flask which contains sucrose.

42. Which of the following statements is correct about the hydrolytic reaction of sucrose? ( )

A. It's the first order reaction, and water doesn't participate in the reaction.

B. It's the fake first order reaction, and water concentration changes little during the reaction.

C. H^+ doesn't participate in the reaction.

D. H^+ concentration changes little during the reaction

43. In the experiment of measuring rate constant for bromination of acetone, which of the following equipments is the best for convenient and accurate measurement ( )

A. conductivity meter B. Abbe refractometer

C. Spectrophotometer D. polarimeter

44. In the experiment of measuring rate constant for bromination of acetone, the concentration calculated using the absorbance is ( )

A. c_{Br_2} B. c_{H^+}

C. $c_{丙酮}$ D. none of the above

45. In the experiment of measuring rate constant for bromination of acetone, the method used to measure the reaction order is ( )

A. half-life method B. integrate method

C. isolated method D. initial rate method

46. In the experiment of saponification of ethyl acetate, physical method can be used to determine the concentration change of ethyl acetate. The selected measuring instrument is ( )

A. polarimeter B. conductivity meter

C. Abbe refractometer D. boiling point meter

47. In the experiment of saponification of ethyl acetate, when half of sodium hydroxide being added to ethyl acetate is taken as the starting point of the reaction, regardless of the volatilization of the ester, the experiment result will show ( )

A. positive error B. negative error

C. no error D. immeasurable error

48. In the experiment of saponification of ethyl acetate, if the ethyl acetate solution is prepared in advance, it will cause the conductivity value ( )

A. to increase B. to decrease C. to changed D. uncertain

49. In the experiment of saponification of ethyl acetate, the conductivity of the system with time ( )

A. increases B. decreases

C. is unchanged D. increases after decreasing

50. In the experiment of determination of rate constant and activation energy for reaction between potassium iodide and hydrogen peroxide, the solution which needs to be added precisely is ( )

A. KI solution B. starch solution

C. $Na_2S_2O_3$ solution D. H_2SO_4 solution

51. Which of the following statements is correct about the experiment of determination of rate constant and activation energy for reaction between potassium iodide and hydrogen peroxide? ( )

A. If timing does not start when first 1ml of $Na_2S_2O_3$ just being consumed, the experiment is failed and redoing is required.

B. Once the solution turns blue, add 1ml of $Na_2S_2O_3$ first and then start the stopwatch.

C. If the reaction rate is too much fast for another run at higher temperature, approach of decreasing $c_{Na_2S_2O_3}$ can be employed to reduce the reaction rate.

D. When preparing KI solution, if a blue solution appears due to free I_2 in iodide solution, a few drops of $Na_2S_2O_3$ can be added to eliminate the bluish appearance.

52. In the experiment of determination of the solution surface tension by bubble pressure method, if the capillary is inserted into the solution, the obtained result will be ( )

A. larger B. smaller C. unchanged D. uncertain

53. In the experiment of determination of the solution surface tension by bubble pressure method, the instrument for accurately determining the concentration of isopentanol (or butyl alcohol) is ( )

A. conductivity meter B. Abbe refractometer

C. spectrophotometer D. polarimeter

54. In the experiment of determination of the solution surface tension by bubble pressure method, the wrong experimental operation is that ( )

A. the capillary must be strictly cleaned

B. the capillary should be vertically placed and the bottom is just tangent to the surface of the liquid surface

C. the capillary nozzle must be smooth and flat

D. the capillary is inserted vertically inside of the liquid and the depth of each insertion keeps the same as possible

55. In the experiment of determination of the solution surface tension by bubble pressure method, the curvature radius r of the bubbles escaping from the capillary ( )

A. becomes larger gradually

C. becomes larger and then smaller gradually

B. becomes smaller gradually

D. becomes smaller and then larger gradually

56. In the experiment of the adsorption of solids in solution, what is the effect of the following operations on the experimental results? ①The conical flask used for titration is not completely dried; ②The conical flask used for titration is rinsed twice with the tested solution. ( )

A. without effect for ①, with effect for ②

B. without effect for ① and ②

C. with effect for ① and ②

D. with effect for ①, without effect for ②

57. In comparison with the experimental value, the specific surface area calculated according to Langmuir monolayer adsorption model is ( )

A. larger

B. smaller

C. approximately equivalent to the experimental value

D. meaningless due to large error

58. In the experiment of adsorption of solids in solution, if a small amount of activated carbon is taken when sampling from the conical flask, the effect on the result is ( )

A. none

B. that the obtained equilibrium concentration will be large, resulting in a larger calculated saturated adsorption amount

C. that the obtained equilibrium concentration will be small, resulting in a larger calculated saturated adsorption amount

D. that the obtained equilibrium concentration will be large, resulting in a smaller calculated saturated adsorption amount

59. Among the following dispersion systems, the one which has the strongest Tyndall effect is ( )

A. the air
B. the sucrose aqueous solution
C. the macromolecular solution
D. the silica sol

60. For the sol prepared by mixing 0.08mol/L KI and 0.1mol/L $AgNO_3$ solutions with equal volume, the order of coagulation capacity for electrolytes $CaCl_2$, Na_2SO_4 and $MgSO_4$ is ( )

A. $Na_2SO_4>CaCl_2>MgSO_4$
B. $MgSO_4>Na_2SO_4>CaCl_2$
C. $Na_2SO_4>MgSO_4>CaCl_2$
D. $Na_2SO_4< MgSO_4< CaCl_2$

61. In the experiment of measuring ζ potential, the HCl solution added into the electrophoresis

tube is ( )

A. concentrated HCl

B. 0.1mol/L HCl

C. HCl with the same conductivity as that of the $Fe(OH)_3$ sol

D. all of the above

62. In the experiment of preparation of sols, $Fe(OH)_3$ sol is prepared by ( )

A. solvent substitution method B. peptization method

C. hydrolytic method D. none of the above

63. In the experiment of determination of average relative molar mass of the macromolecule by viscosity method, the physical quantity measured directly is ( )

A. time B. relative viscosity

C. specific viscosity D. intrinsic viscosity

64. In the experiment of determination of the viscosity of macromolecules, the medium used in the thermostat is ( )

A. silicone oil B. glycerin

C. water D. liquid to be tested

65. Which of the following statements is incorrect about the experiment of determination of average relative molar mass of the macromolecule by viscosity method? ( )

A. Capillary viscometer is used to measure the viscosity of macromolecules. The main instruments and equipment required are capillary viscometer, constant temperature system and stopwatch.

B. The types of viscometers for measuring liquid viscosity include capillary, falling ball and rotating (or torsion) viscometers.

C. In the experiment, the physical quantities directly measured are the time required for the solution to flow through the capillary and the temperature of thermostat.

D. When measuring the viscosity of liquid by capillary method, it is not necessary to use constant temperature system.

Section Two Short Questions

1. In the experiment to determine the heat of combustion, in order to avoid the heat leakage from the systems, what measures need to be taken?

2. Describe the main error sources in the determination of heat of combustion in a bomb calorimeter.

3. What instruments are used in measuring the heat of solution by electrothermic compensation method?

4. What are the factors that the solution heat of substance depends on?

5. A substance with molecular weight of 146 is soluble in cyclohexane, but its existence state (monomer or dimer, or both in equilibrium state) in cyclohexane is unknown. Try to design the experiment to point out the existence state of the substance.

6. In the experiment to determine the saturated vapor pressure of pure liquids by static method, can the system leakage be checked under heating? Why?

7. In the experiment to determine the equilibrium constant of reaction and partition coefficient, which conical flasks should be dried? Which flasks don't need to be dried? Why?

8. In the experiment to determine the equilibrium constant of reaction and partition coefficient, what experimental data do you need to measure directly in the experiment? How to measure it?

9. In the experiment to determine the equilibrium constant of reaction and partition coefficient, why should the flask with plug and grinding mouth be plugged immediately after the solution is installed?

10. In the experiment of phase diagram drawing of completely miscible two-liquid system, how to judge the liquid-vapor equilibrium is reached?

11. What should we pay attention to in the experiment of phase diagram drawing of completely miscible two-liquid system?

12. In the ternary phase diagram experiment, what data need to be precisely recorded to draw the boundary line (that is the binodal solubility curve or double junction line) which divides single-phase and two-phase regions? What data need to be known first and what data need to be calculated?

13. What should we pay attention to when operating the conductometer?

14. In the experiment of determination of thermodynamic functions by electromotive force method, the chemical reaction is $Ag(s) + \frac{1}{2}Hg_2Cl_2(s) \longrightarrow AgCl(s) + Hg(l)$, the cell designed is $Ag(s)|AgCl(s)|KCl(0.1mol/L)|Hg_2Cl_2(s)|Hg(l)$. Please explain why 0.1mol/L KCl but not the high concentrated KCl is used here?

15. In the experiment of measuring the rate constant of sucrose conversion reaction by polarimetry method, when mixing the sucrose solution and the HCl solution, we add the HCl solution to the sucrose solution. Can we mix them the other way round?

16. In the experiment of rate constant determination for saponification of ethyl acetate by conductometric method, why should the experiment be proceeded under the constant temperature? And why should keep the solution in the thermostat in advance before mixing?

17. In the experiment of determining the rate constant of the reaction between potassium iodide and hydrogen peroxide, what is the effect of the acidity and alkalinity of the solution on the experiment?

18. What are the major factors influencing the determination results of rate constant of acetone bromide reaction?

19. What are the factors influencing the determination results of surface tension? How to reduce or eliminate the effect of these factors on the experimental results?

20. In the experiment of adsorption of solids in solution, how to know the adsorption equilibrium is reached?

21. What are the basic conditions for the formation of sol?

22. What factors are related to the speed of electrophoresis?

23. In the experiment of determination of ζ potential for $Fe(OH)_3$ sol, dilute HCl solution is used as the auxiliary liquid. What are the requirements for selecting auxiliary liquid?

24. In the experiment of determination of ζ potential of sol, why should the conductivity of the auxiliary liquid be close to that of the sol as possible?

25. What is the function of branch pipe C in Ubbelohde viscometer? Is it possible to measure the viscosity if pipe C is removed?

26. What effect does the capillary tube of Ubbelohde viscometer have on the measurement result if it is too thick or too thin?

27. Why use intrinsic viscosity $[\eta]$ to calculate the average molecular weight of macromolecules? Is it different from the pure solvent viscosity?

Section Three Answers

Multiple Choice

1. D	2. B	3. A	4. A	5. D	6. B	7. C	8. A	9. A
10. B	11. B	12. D	13. B	14. D	15. C	16. B	17. A	18. A
19. B	20. D	21. C	22. A	23. B	24. B	25. A	26. C	27. B
28. D	29. C	30. A	31. B	32. B	33. C	34. A	35. A	36. D
37. B	38. D	39. A	40. C	41. D	42. B	43. C	44. A	45. D
46. B	47. B	48. A	49. B	50. C	51. D	52. A	53. B	54. D
55. C	56. A	57. B	58. D	59. D	60. C	61. C	62. C	63. A
64. C	65. D							

Short Questions

1. Answer: The measures can be done are: ①to seal the cover; ②to make the inner surface of water bucket polished; ③to use an air layer.

2. Answer: The main error sources include: ① sample weighing; ② heat leakage between system and environment; ③degree of combustion completeness; ④unequal amounts of water used in two experiments; ⑤additional thermal effect caused by agitation; ⑥the thermal effect caused by the change of N_2 in the air into nitric acid; ⑦sensitivity of the instrument for system heating, etc.

3. Answer: The instruments used include direct current supply, digital voltmeter (with an accuracy of 0.001V), digital ammeter (with an accuracy of 0.001A), calorimeter (consisted of thermos cup and electric heater), magnetic stirrer, digital Beckmann thermometer, tablet machine, chronograph, analytical balance and table balance.

4. Answer: The process of dissolution includes the damage of the crystal lattice accompanied with the ionization of solutes and the interaction between the solvent and the solute molecules or ions, which is endothermic reaction for the former and exothermic reaction for the latter. The heat of solution is the sum of the heat involved in all the processes. Therefore, it can be either positive or negative based on the relative contents of solvents and solutes. Besides, the heat of solution also depends on the pressure, temperature and type of the solvent.

5. Answer: Relative molecular weight (or molar mass) can be calculated by determining the freezing point depression: $M_B = K_f W_B / \Delta T_f W_A$. If the freezing point depression constant K_f of solvent is known, T^* of pure cyclohexane and T of the solution having the solvent weight of W_A and solute weight of W_B are determined respectively, the molecular weight of solute (M_B) can be calculated using the equation: $M_B = K_f W_B / \Delta T_f W_A$, where $\Delta T_f = T^* - T$. If M_B equals to 146, the solute exists as a monomer in solution. If M_B is 292, it exists as a dimer in solution. And if M_B is between 146 and 292, both the monomer and dimer exist in equilibrium.

6. Answer: No. Because in the heating process, the temperature is not constant, the gas-liquid phases can't reach equilibrium, leading to non-constant pressure and inaccurate experimental results.

7. Answer: ①The conical flask used for titration with standard solution of $Na_2S_2O_3$ and the flask containing saturated aqueous solution of iodine need not to be dried, because it has no influence on the amount of iodine and will not cause deviation; ②The flask containing KI solution needs to be dried, because the existing water will affect the concentration of iodine ions.

8. Answer: ①Temperature of the thermostat, iodine concentration in water layer and CCl_4 layer. ②The temperature is measured by common thermometer; the iodine concentration in the solution is titrated with sodium thiosulfate standard solution.

9. Answer: Because iodine and carbon tetrachloride are volatile, the grinding plug should be stopped immediately to prevent volatilization.

10. Answer: When thermometer reading basically does not change, or the compositions of the gas and liquid phases basically do not change with time (measured refractive index is basically the same), it can be considered that the gas and liquid phases are basically in equilibrium.

11. Answer: ①The system shouldn't be airtight to ensure the consistency of pressure inside and outside. ②Determine the refractive index of gas condensate before that of liquid condensate. ③The heating rod should be immersed in the solution 2cm or so. ④Do not touch the surface with the pipette to avoid scratching the refracting prism while using the Abbe refractometer. If necessary, gently wipe the prism surface with wiping paper, do not wipe with filter paper.

12. Answer: ①The data need to be correctly recorded include the volume of each component when solution turns from clear to turbid, experimental temperature. ②The datum needs to be known is the density of each component. ③The data need to be calculated are the mass and mass percentage of each component, then plot in the equilateral triangle coordinate.

13. Answer: ①Do not wet the electrode wire otherwise you will get inaccurate data. ②A quick measurement of the conductivity for water is important, otherwise the dissolution of CO_2 in water may lead to a quick increase of the conductivity of water and influence the experimental results. ③All the containers should be clean and with no ion contaminated. ④The platinized surface of the electrode can not be touched when wipe the electrode, or the platinum black will be broken off and result in a variation in cell constant.

14. Answer: Although the EMF of this cell does not relate to the activity of KCl solution, that is $E = E^\ominus$, concentrated KCl may result in dissolution of silver-silver chloride electrode. Therefore,

0.1mol/L KCl solution is selected as the electrode liquid.

15. Answer: No. because H^+ is a catalyst. If we add sucrose to HCl solution, the H^+ concentration is very large. At the beginning of the measurement, most of the sucrose has already reacted to produce fructose and glucose. The optical rotation corresponding to time t is no longer accurate. On the contrary, if HCl solution is added to the sucrose solution, because H^+ concentration is small, the reaction rate is small, and the extent of reaction performed before the timing is small.

16. Answer: The chemical kinetic experiment is usually proceeded under constant temperature because the reaction rate constant depends on the temperature. Besides, the conductivity also depends on the temperature. Therefore, to precisely perform the kinetic experiment, the temperature should be constant. To keep the solution in the thermostat in advance before mixing can reduce the temperature fluctuation, and therefore reduce the experiment error.

17. Answer: In this experiment, the reaction to be studied is $2H^+ + 2I^- + H_2O_2 \longrightarrow 2H_2O + I_2$. Under neutral and alkaline conditions, hydrogen peroxide is catalyzed by potassium iodide to decompose into oxygen, while under acidic condition, hydrogen peroxide is reduced to water, so it must be guaranteed to be acidic in this experiment.

18. Answer: Influencing factors: solution concentration, wavelength, temperature, and reaction time.

19. Answer: Influencing factors: Temperature, the escape velocity of the bubbles, the cleanliness of capillary, and the relative position of capillary nozzle to the surface of the liquid, etc.

Measures to be taken: To keep constant temperature of the tested solution, to control the constant speed of bubbles produced, to keep the capillary strictly clean and to control the capillary nozzle to be tangent to the surface of the liquid, etc.

20. Answer: Take same amount of solution at different time and perform titration, if the volume of NaOH solution used in titration does not change with time, it can be determined that adsorption equilibrium is reached.

21. Answer: The necessary conditions for the formation of hydrophobic sol include ① the solubility of the dispersed phase should be low; ② the existence of a stabilizer, otherwise, the colloidal particles could easily coalesce and agglomerate.

22. Answer: The ζ potential can be calculated using the equation: $\zeta = 9\times10^9 \cdot \dfrac{6\pi\eta v}{\varepsilon_r E}$, then $v = \dfrac{1}{9\times10^9} \cdot \dfrac{\zeta\varepsilon_r E}{6\pi\eta}$. Therefore, the speed of electrophoresis is proportional to the zeta potential of the sol, the dielectric constant ε_r of the medium, and the electric field strength E between the two poles of the electrophoresis apparatus, but it is inversely proportional to the viscosity η of the sol.

23. Answer: ① no chemical reaction with sol; ② no coagulation effect to the sol; ③ to form a clear interface between auxiliary liquid and sol; ④ having possibly the same conductivity as the sol.

24. Answer: ① to guarantee the moving velocity same as that of the sol, and to avoid the

sudden change of electric field strength at the interface which leading to unequal moving velocity and therefore unclear interface; ② Only if both the sol and auxiliary liquid have the similar conductivities is the calculating formula validity.

25. Answer: The function of branch pipe C is to connect with the atmosphere, so that the liquid in pipe B falls solely due to gravity but not to any other factors. The viscosity can still be measured without branch pipe C. However, the volume of standard solution must be identical to that of the solution to be measured. This is because, when the liquid flows down, the pressure difference is related to the height of liquid in pipe A.

26. Answer: If the capillary is too thick, then the flow time is short, increasing the risk of reading error. If the capillary tube is too thin, then blockage appears frequently.

27. Answer: Intrinsic viscosity is defined as the specific concentration viscosity when the concentration of polymer solution approaching to zero. It represents the contribution of a single molecule to the solution viscosity. Its value does not change with the concentration and has a quantitative relationship with the relative molecular weight of the polymer, $[\eta]=KM_\eta^\alpha$. Therefore, the value of intrinsic viscosity is often used to calculate the relative molecular weight. The $[\eta]$ mainly reflects the internal friction between copolymer molecules and solvent molecules in infinite dilution solution, and pure solvent viscosity reflects the internal friction between solvent molecules.

第四部分 附 录
Part Four Appendices

一、常用物理化学实验数据表
(Common Data of Experimental Physical Chemistry)

表 4-1 常用的物理常数*

Table 4-1 Fundamental Physical Constants*

量的名称 (quantity)	量的符号 (symbol)	量的数值(SI) (numerical value)
光在真空中的传播速度 (speed of light in vacuum)	c	$2.997\ 924\ 58\times10^{8}$ m/s
普朗克常数 (Planck constant)	h	$6.626\ 070\ 15\times10^{-34}$ J·s
万有引力常数 (Newtonian constant of gravitation)	G	$6.674\ 30(15)\times10^{-11}$ $m^3/(kg\cdot s^2)$
重力加速度 (acceleration of gravity)	g	9.806 65 m/s^2
基本电荷 (elementary charge)	e	$1.602\ 176\ 634\times10^{-19}$ C
阿伏加德罗常数 (Avogadro constant)	N_A, L	$6.022\ 140\ 76\times10^{23}$ mol^{-1}
电子质量 (electron mass)	m_e	$9.109\ 383\ 701\ 5(28)\times10^{-31}$ kg
质子质量 (proton mass)	m_p	$1.672\ 621\ 923\ 69(51)\times10^{-27}$ kg
中子质量 (neutron mass)	M_n	$1.674\ 927\ 498\ 04(95)\times10^{-27}$ kg
法拉第常数 (Faraday constant)	$F=N_Ae$	$9.648\ 533\ 212\times10^{4}$ C/mol
摩尔气体常数 (molar gas constant)	R	8.314 462 618 J/(mol·K)

续表

量的名称（quantity）	量的符号（symbol）	量的数值（SI）（numerical value）
玻尔兹曼常数（Boltzmann constant）	k	1. 380 649×10^{-23}J/K
真空介电常数（vacuum electric permittivity）	ε_0	8.854 187 812 8(13)×10^{-12}C/(mol · m)
电子荷质比（electron charge to mass quotient）	e/m_e	1. 758 820 010 76(53)×10^{11}C/kg
里德堡常数（Rydberg constant）	R_∞	1. 097 373 156 816 0(21)×10^{7}m^{-1}
玻尔磁子（Bohr magneton）	μ_B	9.274 010 078 3(28)×10^{-24}J/T
玻尔半径（Bohr radius）	$a_0 = a \mid 4\pi R_\infty$	5. 291 772 109 03(80)×10^{-11}m

* 国际科技数据委员会(CODATA)基本物理常数推荐值(2018)

* CODATA internationally recommended 2018 values of the fundamental physical constants.

表 4-2 不同温度下水的黏度（η）和表面张力（σ）

Table 4-2 Viscosity and Surface Tension of Water at Different Temperatures

t/℃	η×10^3/(Pa · s)	σ×10^3/(N/m)	t/℃	η×10^3/(Pa · s)	σ×10^3/(N/m)
0	1. 787	75.64	25	0.890 4	71.97
5	1. 519	74.92	26	0.870 5	71.82
10	1. 307	74.23	27	0.851 3	71.66
11	1. 271	74.07	28	0.832 7	71.50
12	1. 235	73.93	29	0.814 8	71.35
13	1. 202	73.78	30	0.797 5	71.20
14	1. 169	73.64	35	0.719 7	70.38
15	1. 139	73.49	40	0.652 9	69.60
16	1. 109	73.34	45	0.596 0	68.74
17	1. 081	73.19	50	0.546 8	67.94
18	1. 053	73.05	55	0.504 0	67.05
19	1. 027	72.90	60	0.466 5	66.24
20	1. 002	72.75	70	0.404 2	64.47
21	0.977 9	72.59	80	0.354 7	62.67
22	0.954 8	72.44	90	0.314 7	60.82
23	0.932 5	72.28	100	0.281 8	58.91
24	0.911 1	72.13			

表 4-3 不同温度下液体的密度（10^3kg/m^3）

Table 4-3 Density of Various Liquids at Different Temperature（10^3kg/m^3）

t/℃	水（water）	苯（benzene）	甲苯（methyl benzene）	乙醇（alcohol）	氯仿（trichlormethane）	汞（mercury）	醋酸（acetic acid）
0	0.999 842 5		0.886	0.806 25	1. 526	13.595 5	1. 071 8
5	0.999 966 8	—	—	0.802 07	—	13.583 2	1. 066 0
10	0.999 702 6	0.887	0.375	0.797 88	1. 496	13.570 8	1. 060 3
11	0.999 608 1	—	—	0.797 04	—	13.568 4	1. 059 1
12	0.999 500 4	—	—	0.796 20	—	13.565 9	1. 058 0
13	0.999 380 1	—	—	0.795 35	—	13.563 4	1. 056 8
14	0.999 247 4	—	—	0.794 51	—	13.561 0	1. 055 7
15	0.999 102 6	0.883	0.870	0.793 67	1. 486	13.558 5	1. 054 6
16	0.998 946 0	0.882	0.869	0.792 83	1. 484	13.556 1	1. 053 4
17	0.998 777 9	0.882	0.867	0.791 98	1. 482	13.553 6	1. 052 3
18	0.998 598 6	0.881	0.866	0.791 14	1. 480	13.551 2	1. 051 2
19	0.998 408 2	0.880	0.865	0.790 29	1. 478	13.548 7	1. 050 0
20	0.998 207 1	0.870	0.864	0.789 45	1. 476	13.546 2	1. 048 9
21	0.997 995 5	0.879	0.863	0.788 60	1. 474	13.543 8	1. 047 8
22	0.997 773 5	0.878	0.862	0.787 75	1. 472	13.541 3	1. 046 7
23	0.997 541 5	0.877	0.861	0.786 91	1. 471	13.538 9	1. 045 5
24	0.997 299 5	0.876	0.860	0.786 06	1. 469	13.5364	1. 0444
25	0.997 047 9	0.875	0.859	0.785 22	1. 467	13.534 0	1. 043 3
26	0.996 786 7	—	—	0.784 37	—	13.531 5	1. 042 2
27	0.996 516 2	—	—	0.783 52	—	13.529 1	1. 041 0
28	0.996 236 5	—	—	0.782 67	—	13.526 6	1. 039 9
29	0.995 947 8	—	—	0.781 82	—	13.524 2	1. 038 8
30	0.995 650 2	0.869	—	0.780 97	1. 460	13.521 7	1. 037 7
40	0.992 218 7	0.858	—	0.772	1. 451	13.497 3	——
50	0.988 039 3	0.847	—	0.763	1. 433	13.472 9	——
90	0.965 323 0	0.836	—	0.754	1. 411	13.376 2	

表 4-4 不同温度下水的饱和蒸气压

Table 4-4 Saturated Vapor Pressure of Water at Different Temperatures

t/℃	p/kPa	t/℃	p/kPa	t/℃	p/kPa
0	0.612 5	34	5. 320	68	28.56
1	0.656 8	35	5. 623	69	29.83
2	0.705 8	36	5. 942	70	31.16
3	0.758 0	37	6. 275	71	32.52
4	0.813 4	38	6. 625	72	33.95
5	0.872 4	39	6. 992	73	35.43
6	0.935 0	40	7. 376	74	35.96
7	1. 002	41	7. 778	75	38.55
8	1. 073	42	8.200	76	40.19
9	1. 148	43	8.640	77	41.88
10	1. 228	44	9.101	78	43.64
11	1. 312	45	9.584	79	45.47
12	1. 402	46	10.09	80	47.35
13	1. 497	47	10.61	81	49.29
14	1. 598	48	11.16	82	51.32
15	1. 705	49	11.74	83	53.41
16	1. 818	50	12.33	84	55.57
17	1. 937	51	12.96	85	57.81
18	2. 064	52	13.61	86	60.12
19	2. 197	53	14.29	87	62.49
20	2. 338	54	15.00	88	64.94
21	2. 487	55	15.74	89	67.48
22	2. 644	56	16.51	90	70.10
23	2. 809	57	17.31	91	72.80
24	2. 985	58	18.14	92	75.60
25	3. 167	59	19.01	93	78.48
26	3. 361	60	19.92	94	81.45
27	3. 565	61	20.86	95	84.52
28	3. 780	62	21.84	96	87.67
29	4. 006	63	22.85	97	90.94
30	4. 248	64	23.91	98	94.30
31	4. 493	65	25.00	99	97.76
32	4. 755	66	26.14	100	101.30
33	5. 030	67	27.33		

表4-5 不同温度下水和乙醇的折射率*

Table 4-5 Refractive Indices of Water and Alcohol at Different Temperatures*

t/℃	纯水 (water)	99.8%乙醇 (99.8% alcohol)	t/℃	纯水 (water)	99.8%乙醇 (99.8% alcohol)
14	1. 333 48		34	1. 331 36	1. 354 74
15	1. 333 41		36	1. 331 07	1. 353 90
16	1. 333 33	1. 362 10	38	1. 330 79	1. 353 06
18	1. 333 17	1. 361 29	40	1. 330 51	1. 352 22
20	1. 332 99	1. 360 48	42	1. 330 23	1. 351 38
22	1. 332 81	1. 359 67	44	1. 329 92	1. 350 54
24	1. 332 62	1. 358 85	46	1. 329 59	1. 349 69
26	1. 332 41	1. 358 03	48	1. 329 27	1. 348 85
28	1. 332 19	1. 357 21	50	1. 328 94	1. 348 00
30	1. 331 92	1. 356 39	52	1. 328 60	1. 347 15
32	1. 331 64	1. 355 57	54	1. 328 27	1. 346 29

注：* 相对于空气；钠光波长 589.3nm。

* This table gives the refractive index of water under the radiation of sodium light (589.3nm) with respect to air.

表4-6 不同温度下 KCl 的摩尔溶解热 $\Delta_{isol}H_m$ (kJ/mol)*

Table 4-6 Molar Integral Solution Heat of KCl at Different Temperatures (kJ/mol)*

t/℃	$\Delta_{isol}H_m$	t/℃	$\Delta_{isol}H_m$
5	20.941	20	18.297
6	20.740	22	17.995
8	20.338	24	17.702
10	19.979	25	17.556
12	19.623	26	17.414
14	19.276	28	17.138
15	19.100	30	16.874
16	18.933	32	16.615
18	18.602	34	16.372

注：* 1mol KCl 溶于 200mol 水中的积分溶解热。

* The values are 1mol KCl dissolved in 200mol water.

表 4-7 四种浓度 KCl 在不同温度下的电导率 κ（S/cm）

Table 4-7 Conductivities of KCl Solutions at Different Temperatures（S/cm）

t/℃	c/（mol/L）			
	1. 000*	0.100 0	0.020 0	0.010 0
0	0.065 41	0.007 15	0.001 521	0.000 776
5	0.074 14	0.008 22	0.001 752	0.000 896
10	0.083 19	0.009 33	0.001 994	0.001 020
15	0.092 52	0.010 48	0.002 243	0.001 147
16	0.094 41	0.010 72	0.002 294	0.001 173
17	0.096 31	0.010 95	0.002 345	0.001 199
18	0.098 22	0.011 19	0.002 397	0.001 225
19	0.100 14	0.011 43	0.002 449	0.001 251
20	0.102 07	0.011 67	0.002 501	0.001 278
21	0.104 00	0.011 91	0.002 553	0.001 305
22	0.105 94	0.012 15	0.002 606	0.001 332
23	0.107 89	0.012 39	0.002 659	0.001 359
24	0.109 84	0.012 64	0.002 712	0.001 386
25	0.111 80	0.012 88	0.002 765	0.001 413
26	0.113 77	0.013 13	0.002 819	0.001 441
27	0.115 74	0.013 37	0.002 873	0.001 468
28		0.013 62	0.002 927	0.001 496
29		0.013 87	0.002 981	0.001 524
30		0.014 12	0.003 036	0.001 552
35		0.015 39	0.003 312	
36		0.015 64	0.003 368	

注：* 在空气中称取 74.56g KCl，溶于 18℃水中，稀释到 1L，其浓度为 1.000mol/L（密度 1.044 9g/cm^3），再稀释得其他浓度溶液。

* The concentration of liquor will be 1.000mol/L（density is 1.044 9g/cm^3 at 18℃）when 74.56g KCl is dissolved in 1 000ml water, the other concentration liquor will be diluted by it.

表 4-8 298K 时常见离子在无限稀释水溶液中的摩尔电导率 Λ_m^∞（$S \cdot m^2/mol$）

Table 4-8 Ionic conductivity at infinite dilution at 298 K（$S \cdot m^2/mol$）

离子（ion）	$\Lambda_m^\infty \times 10^4$	离子（ion）	$\Lambda_m^\infty \times 10^4$	离子（ion）	$\Lambda_m^\infty \times 10^4$
Ag^+	61.9	F^-	54.4	IO_3^-	40.5
Ba^{2+}	127.8	ClO_3^-	64.4	IO_4^-	54.5
Ca^{2+}	118.4	ClO_4^-	67.9	NO_2^-	71.8
Cu^{2+}	110	CN^-	78	NO_3^-	71.4
Fe^{2+}	108	CO_3^{2-}	144	OH^-	198.6
Fe^{3+}	204	CrO_4^{2-}	170	PO_4^{3-}	207
H^+	349.82	$Fe(CN)_6^{4-}$	444	SCN^-	66
Hg^+	106.12	$Fe(CN)_6^{3-}$	303	SO_3^{2-}	159.8
K^+	73.5	HCO_3^-	44.5	SO_4^{2-}	160
Mg^{2+}	106.12	HS^-	65	Ac^-	40.9
NH_4^+	73.5	HSO_3^-	50	$C_2O_4^{2-}$	148.4
Na^+	50.11	HSO_4^-	50	Br^-	73.1
Zn^{2+}	105.6	I^-	76.8	Cl^-	76.35

二、参考文献（References）

[1] 安正伟,周利兵. 不同厂家阿奇霉素分散片燃烧热检测及质量评价. 科技与创新,2019,9:118-119.

[2] 唐宇,贺福元,邓凯文,等. 中药燃烧焓、信息熵及生物熵的稳定性揭示其成分间的“虹势性”. 中华中医药杂志,2012,27(4):867-873.

[3] 李启泉,贺福元,罗杰英,等. 中药四气燃烧焓数学模型的建立及初步实验研究. 西安交通大学学报(医学版),2009,30(5):624-627,638.

[4] 彭俊军,靳艾平,陈富偈. 物理化学实验. 武汉:华中科技大学出版社,2021.

[5] 李文坡. 物理化学实验. 北京:化学工业出版社,2021.

[6] 吴慧敏. 物理化学实验. 北京:化学工业出版社,2021.

[7] 国家药典委员会. 中华人民共和国药典:2020 年版.四部. 北京:中国医药科技出版社. 2020,85-86.

[8] 徐开俊. 物理化学实验与指导. 2 版. 北京:中国医药科技出版社,2015.

[9] 李森,高静. 物理化学实验教程. 北京:中国医药科技出版社,2019.

[10] 张师愚. 物理化学实验. 北京:中国医药科技出版社,2014.

[11] 罗俊永,杨文智,张晓攀,等. 活性炭在注射剂生产中的潜在风险分析. 中国新药杂志,2019,28(4):404-407.

[12] 杨希琴,陈东,荀哲,等. 可溶性淀粉交联微球的制备及性质考察. 中国药剂学杂

志,2009,7(3):161-170.

[13] 赵喆,王齐放. 表面活性剂临界胶束浓度测定方法的研究进展. 实用药物与临床,2010,13(2):140-144.

[14] 崔黎丽. 物理化学实验指导(双语). 3 版. 北京:人民卫生出版社,2016.

[15] 李三鸣. 物理化学实验. 北京:中国医药科技出版社,2007.

[16] 李云峰,相明辉,于洋,等. 共振光散射法测定复配表面活性剂的临界胶束浓度. 科技通报,2016,32(8):1-4.

[17] ZHANG X W, WU W. Ligand-mediated active targeting for enhanced oral absorption. Drug Discovery Today,2014,19(7):898-904.

[18] GUO Y,MAO X Y,ZHANG J,et al. Oral delivery of lycopene-loaded microemulsion for brain-targeting: preparation, characterization, pharmacokinetic evaluation and tissue distribution. Drug Delivery,2019,26(1):1191-1205.

[19] MO Y L,DU H L,CHEN B L,et al. Quick-responsive polymer-based thermosensitive liposomes for controlled doxorubicin release and chemotherapy. ACS Biomaterials Science & Engineering,2019,5,2316-2329.

[20] JIN Y,LIU Q,ZHOU C H,et al. Intestinal oligopeptide transporter PepT1-targeted polymeric micelles for further enhancing oral absorption of water-insoluble agents. Nanoscale,2019,11:21433-21448.

[21] ZHAO L,ZHOU Y X,GAO Y J,et al. Bovine serum albumin nanoparticles for delivery of tacrolimus to reduce its kidney uptake and functional nephrotoxicity. International Journal of Pharmaceutics,2015,483:180-187.

[22] MOURA R P,PACHECO C,PÊGO A P,et al. Lipid nanocapsules to enhance drug bioavailability to the central nervous system. Journal of Controlled Release,2020,322:390-400.

[23] HAN S D,LI X P,ZHOU C H,et al. Further enhancement in intestinal absorption of paclitaxel by using transferrin-modified paclitaxel nanocrystals. ACS Applied Bio Materials,2020,3:4684-4695.

[24] PACHECO R P,EISMIN R J,COSS C S,et al. Synthesis and characterization of four diastereomers of mono-rhamnolipids. J Am Chem Soc,2017,139:5125-5132.

[25] SCHOLZ N,BEHNKE T,RESCH-GENGER U. Determination of the critical micelle concentration of neutral and ionic surfactants with fluorometry,conductometry,and surface tension-A method comparison. Journal of Fluorescence,2018,28:465-476.

[26] ZHOU Y H,ZHOU C H,ZOU Y,et al. Multi pH-sensitive polymer-drug conjugate mixed micelles for efficient co-delivery of doxorubicin and curcumin to synergistically suppress tumor metastasis. Biomaterials Science,2020,8,5029-5046.

[27] WU S G,LIANG F Q,HU D N,et al. Determining the critical micelle concentration of surfactants by a simple and fast titration method. Analytical Chemistry,2020,92:4259-4265.